Methods in Molecular Biology™

Series Editor
John M. Walker
School of Life Sciences
University of Hertfordshire
Hatfield, Hertfordshire, AL10 9AB, UK

For further volumes:
http://www.springer.com/series/7651

Statistical Methods for Microarray Data Analysis

Methods and Protocols

Edited by

Andrei Y. Yakovlev

School of Medicine and Dentistry, Department of Biostatistics and Computational Biology, University of Rochester, Rochester, NY, USA

Lev Klebanov

Department of Probability and Statistics, Charles University, Prague, Czech Republic

Daniel Gaile

State University of New York at Buffalo, Buffalo, NY, USA

Editors
Andrei Y. Yakovlev
School of Medicine and Dentistry
Department of Biostatistics and Computational Biology
University of Rochester
Rochester, NY, USA

Lev Klebanov
Department of Probability and Statistics
Charles University
Prague, Czech Republic

Daniel Gaile
State University of New York at Buffalo
Buffalo, NY, USA

ISSN 1064-3745 ISSN 1940-6029 (electronic)
ISBN 978-1-60327-336-7 ISBN 978-1-60327-337-4 (eBook)
DOI 10.1007/978-1-60327-337-4
Springer New York Heidelberg Dordrecht London

Library of Congress Control Number: 2012956545

Printed on acid-free paper

Humana Press is a brand of Springer
Springer is part of Springer Science+Business Media (www.springer.com)

Preface

Microarrays for simultaneous measurement of redundancy of RNA species are widely used for fundamental biology research. They are also tested for their use in personalized medicine in disease diagnosis and prognosis. From the point of mathematical statistics, the invention of microarray technology in the mid-1990s allowed the simultaneous monitoring of the expression levels of thousands of genes. Microarrays for simultaneous measurement of redundancy of RNA species are used in fundamental biology as well as in medical research. Microarray may be considered as an observation of very high dimensionality equal to the number of expression levels. There arise some needs to develop new statistical methods to handle the data of such large dimensionality, especially connected to the fact of small number of observations (which is the number of arrays). Because of the small number of observations the standard asymptotic methods of multivariate statistical analysis appear to be inapplicable.

The aim of the book is to familiarize the readers with statistical methods used nowadays in microarray analysis. It is addressed to everybody who is involved or is planning to be involved in statistical data analysis of microarrays, mostly to statisticians but also to biological researchers.

It was impossible to come over all statistical methods published to date in a book. The selection was made on a basis of mathematical correctness of corresponding methods. Some approaches based on intuitive impressions and having no mathematical support were excluded from our consideration. However, the selection criteria were sometimes not scientific. In many cases they reflect personal taste of the editors. Nevertheless, the editors took pains that other valuable methods should be described or mentioned. The editors invite the interested readers to continue their study of the material beyond this book. We are very grateful to all authors for their writing.

Each chapter can be read as a separate entry. The style of the individual writing is essential to show the knowledge and experience of each team that contributed. However, the reader will be guided from microarray technology to statistical problems of corresponding data analysis. Chapters 1 and 2 provide such prolegomena.

In Chapter 3, an introduction to current multiple testing methodology are presented, with the objective of clarifying the methodological issues involved, and hopefully providing the reader with some basis with which to compare and select methods.

Chapter 4 discusses a method of selecting differentially expressed genes based on a newly discovered structure termed as the δ-sequence. Together with the nonparametric empirical Bayes methodology, it leads to dramatic gains in terms of the mean numbers of true and false discoveries, and in the stability of the results of testing. The results of this chapter can be viewed also as a new method of normalization of microarray data.

Chapter 5 is in some sense connected to Chapter 4. It studies different normalization procedures of gene expression levels. Normalization procedures are used for removing systematical variation which affects the measure of expression levels.

Chapter 6 is dedicated to constructing of multivariate prognostic gene signatures with censored survival data. Modern high-throughput technologies allow us to simultaneously measure the expressions of a huge number of candidate predictors, some of which are likely to be associated with survival. One difficult task is to search among an enormous number of potential predictors and to correctly identify most of the important ones, without mistakenly identifying too many spurious associations. Mere variable selection is insufficient, however, for the information from the multiple predictors must be intelligently combined and calibrated to form the final composite predictor. Many commonly used procedures overfit the training data, miss many important predictors, or both. Author proposes a method that offers a middle ground where some partial multivariate adjustments can be made in an adaptive fashion, regardless of the number of candidate predictors. He demonstrates the performance of our proposed procedure in a simulation study within the Cox proportional hazards regression framework, and applies this new method to a publicly available data set to construct a novel prognostic gene signature for breast cancer survival.

Chapter 7 considers clustering problems for gene-expression data.

There are two distinct but related clustering problems with microarray data. One problem concerns the clustering of the tissue samples (gene signatures) on the basis of the genes; the other concerns the clustering of the genes on the basis of the tissues (gene profiles). The clusters of tissues so obtained in the first problem can play a useful role in the discovery and understanding of new subclasses of diseases. The clusters of genes obtained in the second problem can be used to search for genetic pathways or groups of genes that might be regulated together. Authors focus here on mixtures of normals to provide a model-based clustering of tissue samples (gene signatures) and of gene profiles.

Network-based analysis of multivariate gene expression data is given in Chapter 8. Such data are collected to study genomic responses under special conditions for which the expression levels of given genes are expected to be dependent. One important question from such multivariate gene expression experiments is to identify genes that show different expression patterns over treatment dosages or over time; these genes can also point to the pathways that are perturbed during a given biological process. Several empirical Bayes approaches have been developed for identifying the differentially expressed genes in order to account for the parallel structure of the data and to borrow information across all the genes.

In Chapter 9, author discusses the statistical problem, termed oncogene outlier detection, and discusses a variety of proposals to this problem. A statistical model in the multiclass situation is described; links with multiple testing concepts are established. Some new nonparametric procedures are described and compared to existing methods using simulation studies.

Data quality is intrinsically influenced by design, technical, and analytical parameters. Quality parameters have not yet been well defined for gene expression analysis by microarrays, though ad interim, following recommended good experimental practice guidelines should ensure generation of reliable and reproducible data. In Chapter 10 author summarizes essential practical recommendations for experimental design, technical considerations, feature annotation issues, and standardization efforts.

Inferring gene regulatory networks from microarray data has become a popular activity in recent years, resulting in an ever increasing volume of publications. There are many pitfalls in network analysis that remain either unnoticed or scantily understood. A critical discussion of such pitfalls is long overdue. Chapter 11 discuss one feature of microarray data the investigators need to be aware of when embarking on a study of putative associations between elements of networks and pathways.

Finally, Chapter 12 considers the problem of normality of logs of gene expression levels. In the literature there is no unique point on the fact of normality (or nonnormality) of the distribution of gene expression levels. This chapter discusses different approaches to testing of normality in this situation.

The editors and the contributors assume from the reader a basic knowledge of biological concepts of gene expression and statistical methods of gene expression analysis.

Editors are grateful to all contributors.

Prague, Czech Republic ***Lev Klebanov***

Contents

Contributors

ANTHONY ALMUDEVAR • *Department of Biostatistics and Computational Biology, University of Rochester, Rochester, NY, USA*

GARRETT BRODEUR • *Department of Pediatrics, Children's Hospital of Philadelphia, Philadelphia, PA, USA*

PETER BUBELÍNY • *Department of Probability and Statistics, Charles University, Prague, Czech Republic*

LINLIN CHEN • *School of Mathematical Sciences, Rochester Institute of Technology, Rochester, NY, USA*

MANHONG DAI • *Psychiatry Department and Molecular and Behavioral Neuroscience Institute, University of Michigan, University of Michigan, Ann Arbor, MI, USA*

L.K. FLACK • *Department of Rheumatology, University of NSW, Ryde, NSW, Australia*

DEBASHIS GHOSH • *Departments of Statistics and Public Health Sciences, Penn State University, DuBios, PA, USA*

DIANA M. KELMANSKY • *Instituto de Cálculo, Ciudad Universitaria, Buenos Aires, Argentina*

LEV KLEBANOV • *Department of Probability and Statistics Charles University Prague, Czech Republic*

HONGZHE LI • *Department of Biostatistics and Epidemiology, University of Pennsylvania School of Medicine, Philadelphia, PA, USA*

Department of Pediatrics, Children's Hospital of Philadelphia, Philadelphia, PA, USA

G.J. MCLACHLAN • *Department of Mathematics, University of Queensland, Brisbane, Australia*

FAN MENG • *Psychiatry Department and Molecular and Behavioral Neuroscience Institute, University of Michigan, University of Michigan, Ann Arbour, MI, USA*

JANE MINTURN • *Department of Pediatrics, Children's Hospital of Philadelphia, Philadelphia, PA, USA*

S.K. NG • *School of Medicine, Griffith University, Meadowbrook, Australia*

DERICK R. PETERSON • *Department of Biostatistics and Computational Biology, University of Rochester, Rochester, NY, USA*

XING QIU • *Department of Biostatistics and Computational Biology, University of Rochester, Rochester, NY, USA*

ERIC RAPPAPORT • *Children's Hospital of Philadelphia, Philadelphia, PA, USA*

ANDREAS SCHERER • *Genomics, Biomarker Development, Spheromics, Kontiolahti, Joensuu, Finland*

BOBOSHARIF SHOKIROV • *Department of Probability and Statistics, MFF, Charles University, Prague, Czech Republic*

K. WANG • *University of Queensland, Melbourne, Australia*

STEPHEN WELLE • *Functional Genomics Center, University of Rochester, Rochester, NY, USA*

WEI ZHI • *Department of Biostatistics and Epidemiology, New Jersey Institute of Technology, Newark, NJ, USA*

Chapter 1

What Statisticians Should Know About Microarray Gene Expression Technology

Stephen Welle

Abstract

This chapter briefly reviews how laboratories generate microarray data. This information may give data analysts a better appreciation of the technical sources of variability in the data and the importance of minimizing such variability by normalization methods or exclusion of aberrant arrays.

Key words: DNA arrays, cDNAs, RNAs, rRNAs, mRNAs, hybridization Target quantitation, Basic data processing

1. Introduction

Those who dispense advice about how to analyze gene expression data generated with DNA arrays should understand some of the basics regarding the physical characteristics of DNA arrays, the labeling of nucleic acids with reporter molecules, nucleic acid hybridization, and how the reporter molecules are quantified. This information can help the analyst understand potential limitations of the data, sources of analytical variance, and the language of the biologists and technical personnel who provide the data.

Microarrays are used not only to determine patterns of gene expression but also to examine variations in genomic DNA (e.g., "SNP" chips) and binding of transcription factors to DNA. However, this chapter is limited to a discussion of arrays used to quantify levels of gene transcripts (mRNAs).

Andrei Y. Yakovlev et al. (eds.), *Statistical Methods for Microarray Data Analysis: Methods and Protocols*, Methods in Molecular Biology, vol. 972, DOI 10.1007/978-1-60327-337-4_1, © Springer Science+Business Media New York 2013

2. Characteristics of DNA Arrays

DNA arrays are glass slides, other solid supports, or flexible membranes onto which different species of DNA have been attached in a known pattern. The various DNA species, known as "probes," differ in their base sequences. Each probe is designed to capture a particular "target." The targets are gene transcripts (mRNAs) or molecules of DNA or RNA complementary to the gene transcripts (cDNA or cRNA) that have been labeled with reporter molecules (typically fluorescent dyes) to reveal the relative abundances of targets captured at specific loci on the array. In the early days of DNA arrays, probes were spotted onto the solid support by hand or by robots in individual labs. Spot size, shape, and DNA density were often variable, but this was not considered a major problem because generally they were used for competitive hybridization between two RNA samples ("two color" method described below). Currently the use of commercial oligonucleotide (25–60 bases in length) arrays is much more common. With these arrays, usually referred to as "microarrays," the term "feature" often is used instead of the term "spot." The manufacture of these arrays often involves synthesis of the oligonucleotides directly on the array rather than spotting premade molecules onto the surface. These commercial microarrays have uniform features so that different arrays can be compared to one another using the "one-color" approach in which only a single sample is applied to an array.

High-density oligonucleotide arrays can have more than a million features, allowing for multiple features per target. Each feature has a diameter of only a few microns on very dense arrays, up to about 100 μm on less densely packed arrays (Fig. 1). It is important that the number of probe molecules per feature is not a limiting factor in capturing targets. If it were, then all abundant mRNA species would appear to have the same concentration in all samples. Even small features have millions of probe molecules. Although there can be billions of copies of specific target molecules in the hybridization solution, during a typical overnight hybridization the number of target molecules is the main factor that limits signal intensity because of the inefficiency of hybridization on microarrays.

Criticisms of DNA microarray technology that are based on homemade cDNA arrays (1) should not be generalized to current commercial microarrays. Although a very small percentage of the features may be defective on commercial arrays, replicate arrays for the same sample generally are not necessary. Biological replicates usually are a better use of money and time than technical replicates (2). However, it is important to inspect every individual microarray for obvious flaws such as areas of no (or weak) hybridization,

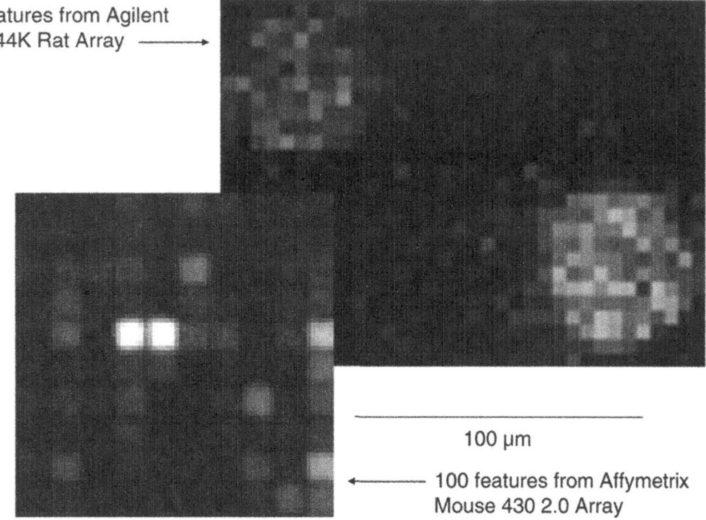

Fig. 1. Magnified images of microarray fluorescence at the level of individual pixels. Only small proportions of the array surfaces are shown. The magnification factor is the same for both arrays. The larger pixels for the Agilent array reflect the use of a scanner with less spatial resolution (5 μm) than the Affymetrix scanner. The lack of visible fluorescence from many of the features is typical of eukaryotic samples, but the usual algorithms will assign a nonzero expression level for these features.

areas with unusually intense hybridization, large scratches, and high background. Inclusion of such arrays can greatly reduce the power to discover differentially expressed genes or correlations among subjects or genes.

3. Preparation of Targets for Hybridization with Microarrays

RNA from biological samples is a complex mixture of ribosomal RNAs (rRNAs), messenger RNAs (mRNAs), pre-messenger RNAs (messenger RNA before it has been spliced and polyadenylated), transfer RNAs (tRNAs), and a number of small RNAs with various functions. Most DNA arrays are designed to measure the pattern of expression of mRNAs, although recently there has been great interest in probing for microRNAs (miRNAs) because they can influence mRNA translation into proteins. Biologists want to know the concentrations of specific mRNA species in relation to the total pool of all mRNAs. This is because mRNA translation into proteins is viewed as a stochastic process in which mRNA molecules are in competition with one another for translation by the protein synthesis machinery of the cells. Of course, this scheme is oversimplified as there are mechanisms for regulating translation of specific mRNAs. But it still makes more sense to determine the concentration of a specific mRNA in relation to the sum of all mRNAs rather than in relation to some "housekeeping gene" that is not necessarily expressed at a consistent level under all experimental conditions or even at the same level in all subjects within the same experimental group.

RNA is extracted by disruption of cell membranes with lysis buffers, with mechanical disruption with homogenizers or pulverization methods also necessary for tissue samples. RNA-degrading enzymes (RNases) are ubiquitous contaminants that can degrade RNA during extraction and storage. Moreover, chemical treatments can cause RNA degradation. Partial RNA degradation is the rule in histology specimens that have been fixed in formalin and embedded in paraffin. Most labs check for RNA degradation by separating different RNA species by size (with electrophoresis) to confirm that most of the RNA is in two major ribosomal species (known as 28S rRNA and 18S rRNA in eukaryotes and 23S and 16S in prokaryotes). If RNA is partially degraded, the prominent rRNA "peaks" or "bands" comprise relatively less of the total RNA (Fig. 2). Discrete rRNA peaks might not be seen at all if the RNA is thoroughly degraded. Low quality RNA samples should not be used for DNA array experiments. There is no standard method for defining whether an RNA sample is degraded too much to use in the experiment. If the gene expression data for a particular sample are unusual compared with other samples collected under the same conditions, the data analyst should consider the possibility that degraded RNA is the cause. If the array work was performed in a competent lab, there should be a record of RNA quality. Normalization methods cannot overcome the problem of poor-quality RNA.

With eukaryotic samples, the mRNA can be harvested from the total RNA by capturing it with beads attached to deoxythymidine oligonucleotides (oligo-dT), which bind to the long (50–200 bases) poly-A tails on mRNAs. However, this step is not necessary with most current protocols for target preparation. The amount of RNA or mRNA that has been extracted from the biological material usually is quantified by measuring the absorbance of ultraviolet (UV) light at 260 nm wavelength. There are certain problems with this method (e.g., other molecules also absorb UV light) that limit its accuracy. Fluorometric methods are more accurate for determining the amount of RNA extracted from a sample but are not used as commonly. Thus usually there is some small variability in the starting amount of RNA because of the imprecision of RNA measurements. Normalization methods should minimize the impact of such variability.

Unless the mRNA is directly labeled with reporter molecules, it is converted to cDNA by an enzymatic process known as reverse transcription (Fig. 3). This cDNA can be labeled during its synthesis, labeled after synthesis, or converted back to RNA by

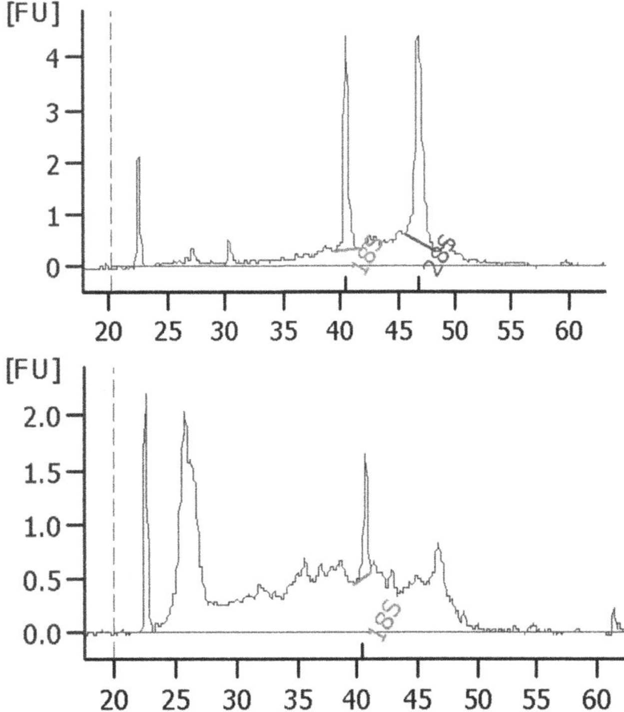

Fig. 2. Assessment of RNA quality with the Agilent Bioanalyzer. The horizontal axis is time (s) and the vertical axis is fluorescence. Larger RNA species take longer to travel through the channels in the Bioanalyzer chips and are detected later. Good samples, such as the one in the *top panel*, have a stronger signal for 28S rRNA (47 s) than for 18S rRNA (40.5 s), and have much stronger signals for rRNA than for mRNA (spread across the entire time scale). Partially degraded samples, such as the one in the *bottom panel*, have lower 28S/18S ratios and much more fluorescence to the left of the 18S rRNA peak. The spike at 23 s is an instrument standard.

in vitro transcription (IVT). The cDNA is "amplified" by IVT—each molecule of cDNA is copied many times. IVT produces complementary RNAs (cRNAs), also known as antisense RNAs (aRNAs). The amplified cRNA population produced by IVT represents the original mRNA population very well. There also is a method to produce hundreds to thousands of copies of cDNA from each original cDNA molecule (3). These amplification methods have enabled labs to do microarray experiments with 5 ng or less of total RNA.

Because rRNAs are the most abundant RNA species by far, they could interfere with the detection of other targets even when the probes have a weak affinity for rRNA sequences. In eukaryotes, almost every mRNA has a poly-A tail whereas other types of RNA do not. Thus, if cDNA is made using a primer (oligo-dT) that pairs with this poly-A tail to start the reaction, the mRNA molecules serve as templates for cDNA synthesis, but the rRNAs do not (Fig. 3). This

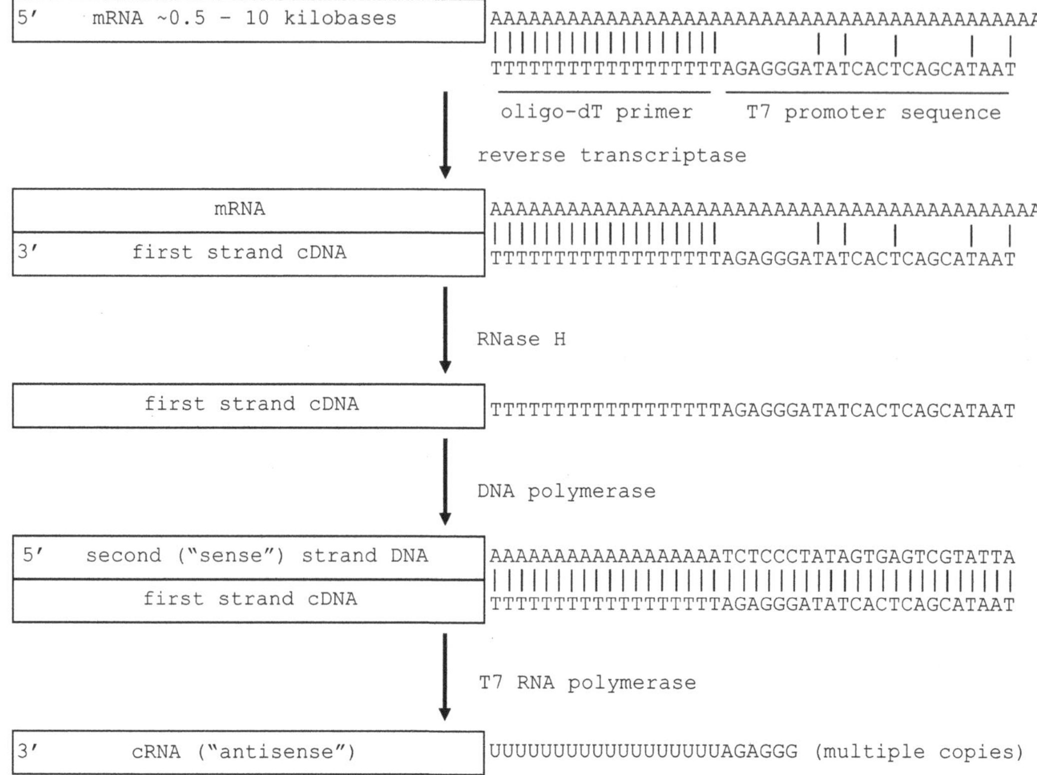

Fig. 3. Synthesis of cDNA and cRNA from eukaryotic mRNA. Reverse transcription requires a primer that hybridizes with the mRNA, and the fact that mRNAs have a long poly-A tail (longer than shown here) allows one to use oligo-dT to initiate reverse transcription of all mRNAs with one primer. The primer shown here also adds a promoter sequence that permits transcription of the double-stranded DNA generated from the mRNA template. The RNA polymerase transcribes each DNA molecule many times so that each original mRNA molecule is represented by many cRNA molecules.

step eliminates the problem of low-affinity rRNA hybridization to the probes. Although low-affinity targets are mostly removed in the washing step following hybridization, it is desirable to minimize rRNAs during hybridization so that they do not hinder interactions of probes with their intended targets. Prokaryotes do not add poly-A tails to their mRNAs, so the best one can do is to remove most of the rRNA by capture with specific oligonucleotides.

The method used to prepare mRNA or cDNA targets depends on the orientation of the DNA probes on the arrays. If the probe base sequences are the same as the mRNA base sequences ("sense" probes), then the targets must be cRNAs or cDNAs. If the probe sequences are reverse complements of the mRNA sequences ("anti-sense" probes), then the targets must be the original mRNAs or second-strand cDNAs (i.e., reverse complements of the cDNA generated by reverse transcription).

In order to detect how many target molecules were captured by an array feature, the targets must be labeled with a reporter molecule (or a molecule that can capture a reporter) because there is no feasible way to detect the capture of ordinary (unlabeled) nucleic acids. There are a number of ways to label the targets, but a discussion of these is beyond the scope of this chapter. The labels are attached covalently so that they cannot be lost during the various hybridization and washing steps. The most common reporters are fluorescent molecules. The labeling efficiency can vary from sample to sample, particularly if the samples are processed in different batches. Thus, while the total amount of mRNA, cRNA, or cDNA placed into the hybridization solution usually is kept constant (within the limits of error of nucleic acid quantitation), less efficiently labeled samples result in dimmer features even though the total number of target molecules is not diminished.

The original mRNA, cDNA, or cRNA molecules are several hundred to a few thousand bases in length. Shorter molecules hybridize with the probes more efficiently. Therefore the target molecules generally are fragmented into smaller pieces before hybridization. Any variability in this fragmentation process can add to variability of hybridization efficiency and add technical noise to the experiment.

4. Hybridization

Hybridization is the binding of a molecule of single-stranded RNA or DNA with a complementary strand. RNA–RNA, DNA–DNA, and DNA–RNA hybrids can form. In microarray experiments, the hybrids of interest are those formed between the DNA probes attached to the array surface and the labeled mRNA, cRNA, or cDNA target molecules generated from the biological sample. The target molecules are placed in a closed chamber in a small volume of fluid that covers the surface of the array, and the chamber is rotated or shaken (usually overnight) to promote movement of the fluid over the surface. The probability that a target molecule will form a stable hybrid with a probe molecule depends on time, rate of movement of target molecules over the surface of the array, temperature, chemical composition of the hybridization solution, and thermodynamic stability of the target–probe hybrid. The last factor is determined by the probe design, with longer and more GC-rich probes producing more stable hybrids. Most manufacturers use the same length for all probes (e.g., Affymetrix uses 25-base oligos and Agilent uses 60-base oligos) regardless of base composition, and therefore the various target–probe hybrids have variable stability. This means that different probes for the same target do not necessarily capture a similar number of target molecules.

A problem with hybridization-based measures of gene expression is that there often is enough complementarity between probes and partially matching target molecules that they hybridize at least transiently. This "cross-hybridization" hinders the target molecule from finding its perfect partner, and also blocks the probe molecule so that its perfect partner cannot hybridize with it. Longer molecules are more problematic in this regard. If the imperfect hybrids are strong enough, a certain percentage will be retained even after washing the arrays, causing an erroneous estimate of the abundance of the perfect-match target. We encountered this problem when studying transgenic mice that express an mRNA with a long CUG-repeat tract. Probes with a few CTG sequences (complementary to the CUGs) yielded high signals, even though expression levels of the intended target transcripts were not increased. Another example of the cross-hybridization problem is that probes for Y-chromosome transcripts yield signals when hybridized with cRNA generated from female subjects. One can only hope that the extent of cross hybridization for most probes is small enough that it does not have a significant impact on quantifying the intended targets.

Hybridization of targets to probe molecules that are immobilized on a solid surface is far less efficient than hybridization when both target and probe molecules are in solution. After a typical overnight hybridization, only a small percentage of the target molecules has been captured by probes. For example, when we hybridized labeled cDNA with one array on day 1, then used the same cDNA solution to hybridize to a new array of the same type on day 2, we found that the average raw signal intensity was only 11% lower on day 2. This inefficient hybridization limits sensitivity, but an advantage is that the targets usually do not saturate the features even when there are many more target molecules than probe molecules. Although the features generally are not completely saturated by the abundant targets, the fraction of target molecules hybridizing to the probes tends to decline at high target concentrations (Fig. 4).

The standard protocols for using the popular Affymetrix microarrays involve using a separate microarray for every sample, a "one color" approach. Some labs prefer the "two color" approach, which involves labeling one sample with a fluorescent dye that emits light at one wavelength and labeling another sample with a dye that emits light at a different wavelength. "Cy5" and "Cy3" are commonly used reporter dyes for this method. Both samples are hybridized with the same array so that the ratio of one color to the other reflects the relative expression level in the two samples. With this approach, variability in spot size and probe density should not have a significant impact because only the ratio of signals, not the total signal intensity, determines the outcome. Some normalization of the ratios is necessary because it is difficult to match the total target concentration precisely and because there are biases associated with

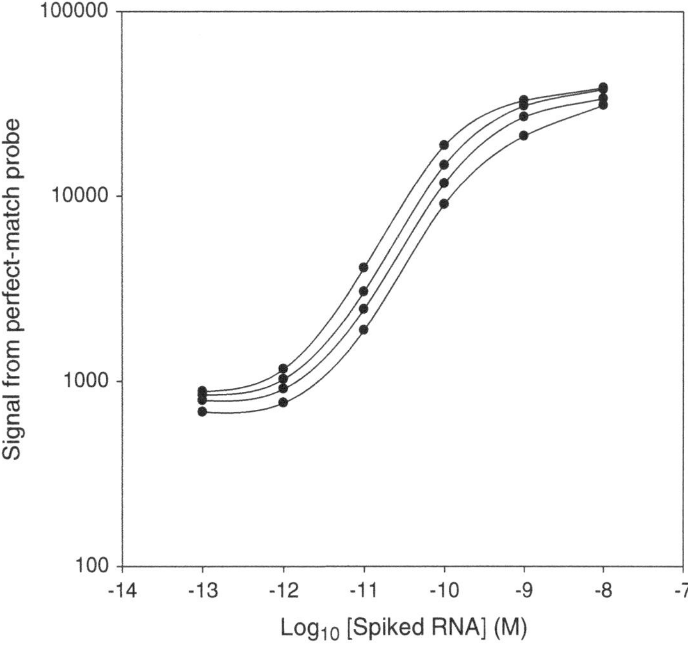

Fig. 4. Relation between signal intensities and molar concentrations of four different bacterial RNAs spiked into eukaryotic RNA. Affymetrix arrays have probes designed to capture these bacterial RNAs. Note the nonlinearity at low and high RNA concentrations. Perfect-match (PM) data from Table 1 of Chudin et al. (8).

the different fluorescent dyes. The Microarray Quality Control study indicated that data quality is similar with the one color and two color approaches (4). I favor the one color approach.

5. Post-hybridization Washing and Target Quantitation

After the hybridization period, the arrays must be washed to remove targets that are only weakly bound to features with partial complementarity. Washing conditions must be established so that hybrids with perfect complementarity are stable during the washing. Overly stringent washing reduces sensitivity by stripping perfect-match targets from their probes.

If the targets have been labeled with a dye, then after washing they are ready for scanning. With some systems the targets in the hybridization solution are not directly labeled with a fluorescent dye. Instead, they are attached to molecules that capture the fluorescent molecules after hybridization has been completed. For example, the standard Affymetrix protocol employs biotinylated targets. The biotin captures streptavidin-phycoerythrin

(SAPE), which is fluorescent, in a "staining" step after washing. Usually, the signal is amplified by using an anti-streptavidin antibody conjugated to multiple biotins, which in turn capture more SAPE molecules in a second round of staining. Just as the efficiency of labeling with a fluorescent dye, biotin, or other molecules influences the final signal generated per hybridization event, so does the efficiency of staining with SAPE or antibodies.

The fluorescence intensity at each feature is detected with a microarray scanner. As it scans along the array surface, this device emits a laser beam with a wavelength appropriate for exciting the fluorescent reporter molecules. The amount of higher-wavelength light emitted from the surface, which is proportional to the number of reporter molecules excited by the laser beam, is recorded simultaneously. The spatial resolution of scanners generally is 5 μm or less, smaller than the feature diameter, so that a feature is comprised of numerous pixels (Fig. 1). Typically, 16 bit analog to digital converters are used for light quantitation and the image files have signal intensities for each pixel represented as integers of $0–2^{16}$. There is signal amplification in converting photons to electrons (i.e., relative voltages recorded in the image file). Generally a lab keeps the signal amplification constant across experiments. If it is set too high there can be truncation of strong signals. But if it is set too low the sensitivity to detect less abundant targets is reduced.

6. Basic Data Processing (Pixel Averaging, Background Subtraction, Normalization)

The statistician may receive raw microarray images or raw signal intensity data, but often the data presented to the statistician already have been processed to some extent. Before a microarray is used for higher level statistical analysis, someone should look its raw image to ensure that there are no significant flaws. Ideally, the lab that processed the microarrays should have done this, but if it cannot be confirmed that such inspections were done it is worth the effort to do so.

The array scanners typically have software that is used by the lab to average the signals from the multiple pixels that comprise each feature, and the statistician rarely has to be concerned about doing this step. Some programs provide a measure of the variability, e.g., the standard deviation, of the individual pixels that were used to compute the reported signal intensity.

Background can be defined either as the mean signal where there is no DNA attached to the array (due to the positive voltage bias of the scanner) or as the mean signal of DNA features for which there is no target. The latter is a more appropriate background as it accounts for nonspecific hybridization signals, but of course this value varies among features depending on the GC content of the probe and the number and abundance of targets that can cross hybridize. Affymetrix software uses the lowest 2% of signals to define the background. Many arrays include features that do not match any known mRNA sequence in order to define the background. "Fold-change" values are underestimated when there has been no background subtraction. The reported background signal values should be fairly consistent across all arrays in an experiment. A small error in the background correction can have a large effect on the computed expression levels for transcripts expressed at low levels. The data analyst should consider excluding any arrays that are clear outliers with respect to background intensity.

Ceiling effects can be a problem. Complete saturation of features because of too many target molecules in the hybridization solution generally does not occur, but linearity between target concentration and signal intensity suffers at high target concentrations (Fig. 4). Moreover, it is not uncommon to have truncation of a signal because of a high gain setting on the scanner. Truncation renders data for the most abundant transcripts useless. If there are too many truncated features on an array, normalization (see below) could be significantly affected. Truncation is detected by plotting signals from pre-normalized arrays against one another (Fig. 5).

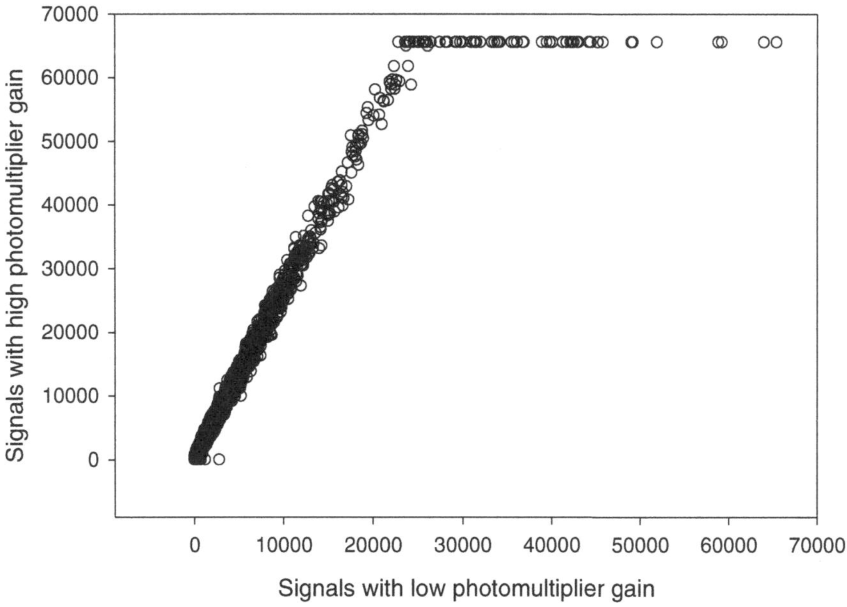

Fig. 5. Signal truncation at high photomultiplier gain setting. Arrays hybridized with cRNA from rat kidneys were scanned at two different multiplier settings. When gain was lowered to avoid truncation, only two signals approached the maximum value (65,536). When the gain was raised to improve detection of weak signals, there was a close correlation with values from the low-gain scan, but many signals reached the maximum value and it is clear from this plot that these values are too low because of truncation.

Normalization refers to the process whereby the signals are adjusted so that the sum of signal intensities is uniform for all arrays in an experiment (or the mean ratio is 1 in a 2-color experiment). In my opinion, normalization always should be done. Even though stringent normalization methods (e.g., quantile normalization) might suppress some interesting biological variance, a greater danger is not adequately accounting for the variance caused by technical factors as described throughout this chapter. A simple method for normalization is to multiply all signals from a particular array by a scaling factor such that the mean signal across all features is the same for all arrays in the experiment (Affymetrix software excludes the highest 2% and lowest 2% of signals from this calculation). This approach ensures that the signal for every transcript is expressed in relation to the total mRNA, which is the value that biologists want to know. More stringent normalization protocols, such as lowess or quantile normalizations (5, 6), are useful when there are intensity-dependent deviations among arrays.

In principle, two-color experiments are self-normalizing because the measure of interest is the ratio between two different colors at the same feature. However, in practice there must be normalization for dye biases and intensity-dependent variations in color ratios.

Statisticians should consider elimination of data from noninformative features before proceeding with the higher level analysis. Anyone who looks at microarray image files appreciates the fact that a significant percentage of the features are as dim as the background (Fig. 1). For many features, the signal is at the background level for all subjects in the experiment, because no tissue or cell expresses every gene. However, expression does not get scored as "zero" because algorithms have been developed to avoid expression scores of zero to facilitate use of logarithmic data. Inclusion of these uninformative features reduces power because *P* values and false discovery statistics depend on the number of features included in the analysis. Probes with very weak signals for all subjects contribute a disproportionate number of false positives in experiments designed to detect differential expression (7). I include only those features that represent targets that are actually present in at least half of the samples from the experimental condition with the highest mean expression level. This might remove more than half of the features in some experiments. Defining which targets are present is somewhat arbitrary. For Affymetrix arrays, I use a nonconservative level ($P < 0.10$) of their "detection *P*" value (this is based on a comparison of perfect-match with single-base-mismatch probes) to define an adequate level of target-specific signal, so as not to eliminate too many features. For platforms that do not generate detection *P* or a similar metric for individual features, a signal intensity 2–3 standard deviations above that of the very dimmest features might be used as a cutoff.

References

1. Lee ML, Kuo FC, Whitmore GA, Sklar J (2000) Importance of replication in microarray gene expression studies: statistical methods and evidence from repetitive cDNA hybridizations. Proc Natl Acad Sci USA 97(18):9834–9839
2. Han ES, Wu Y, McCarter R, Nelson JF, Richardson A, Hilsenbeck SG (2004) Reproducibility, sources of variability, pooling, and sample size: important considerations for the design of high-density oligonucleotide array experiments. J Gerontol A Biol Sci Med Sci 59(4):306–315
3. Dafforn A, Chen P, Deng G, Herrler M, Iglehart D, Koritala S et al (2004) Linear mRNA amplification from as little as 5 ng total RNA for global gene expression analysis. BioTechniques 37(5):854–857
4. Patterson TA, Lobenhofer EK, Fulmer-Smentek SB, Collins PJ, Chu TM, Bao W et al (2006) Performance comparison of one-color and two-color platforms within the MicroArray Quality Control (MAQC) project. Nat Biotechnol 24(9):1140–1150
5. Quackenbush J (2002) Microarray data normalization and transformation. Nat Genet 32(Suppl):496–501
6. Bolstad BM, Irizarry RA, Astrand M, Speed TP (2003) A comparison of normalization methods for high density oligonucleotide array data based on variance and bias. Bioinformatics 19:185–193
7. McClintick JN, Edenberg HJ (2006) Effects of filtering by present call on analysis of microarray experiments. BMC Bioinform 7:49
8. Chudin E, Walker R, Kosaka A, Wu SX, Rabert D, Chang T et al (2001) Assessment of the relationship between signal intensities and transcript concentrations for Affymetrix GeneChip arrays. Genome Biol 3:research0005.1–research0005.10

Chapter 2

Where Statistics and Molecular Microarray Experiments Biology Meet

Diana M. Kelmansky

Abstract

This review chapter presents a statistical point of view to microarray experiments with the purpose of understanding the apparent contradictions that often appear in relation to their results. We give a brief introduction of molecular biology for nonspecialists. We describe microarray experiments from their construction and the biological principles the experiments rely on, to data acquisition and analysis. The role of epidemiological approaches and sample size considerations are also discussed.

Key words: Microarray experiments, Image processing, Calibration, Statistics, Epidemiology

1. Introduction

This chapter is written for statisticians that are faced with the challenge of getting into the increasing area of genomics and for biologists who find that is difficult to interact with statisticians.

The first difficulty that statisticians and biologists encounter is the achievement of a common interdisciplinary language. This means understanding new words with new meanings and old words with different meanings and being open to having no strictly defined concepts. Gene is a good example of a concept in the process of evolving. From classical genetics its meaning rooted in the Mendelian model of monogenic diseases "the gene for": the gene for breast cancer, the gene for hypercholesterolemia, the gene for schizophrenia. However "the gene for" is rather the gene modification that increases the odds of a person to get a certain disease and moreover it is now known that genes act in

Andrei Y. Yakovlev et al. (eds.), *Statistical Methods for Microarray Data Analysis: Methods and Protocols*, Methods in Molecular Biology, vol. 972, DOI 10.1007/978-1-60327-337-4_2, © Springer Science+Business Media New York 2013

mutual coordination with one another and with the environment. The term gene is used with different semantics by the major international genomic databases (1). It was originally described as a "unit of inheritance" and it has derived to a "set of features on the genome that can produce a functional unit."

The genome of any kind of organism, including humans, is the complete information needed to build and maintain a living specimen of that organism. This information is encoded in its *deoxyribonucleic acid* (*DNA*) and ranges from a few million nucleotides for a bacterium or a few billion nucleotides for a eukaryote. Every cell of our body contains the same genetic information, but what makes the unique properties of each cell type? Only a fraction of this information is active in what is called "gene expression."

Microarrays technologies provide biologists with indirect measures of the abundance of thousands of expressed DNA sequences (cDNA) or the presence of thousands of DNA sequences in an organisms' genome.

Statistical scientists might be wondering what terms like DNA, cDNA, nucleotides, genes, genome, gene expression, and eukaryote mean and what microarray technologies are.

We will begin with a brief review of molecular biology to familiarize a statistical reader with many genomic terms that are frequently encountered in relation to microarray experiments. Also we will present statistical points of view that may help biologists towards a deeper insight of their experiments random aspects.

2. A Brief Introduction to Molecular Biology

Microarray experiments are usually trying to identify genes with different expression levels (differentially expressed) among several conditions. We will present the relevant biological concepts and at the end of this section a statistical reader should understand the phrase "gene expression level."

2.1. Nucleic Acids (DNA–RNA)

Nucleic acids can be classified in two types:

DNA, usually presenting a *double* stranded *nucleotide chain* structure.

RNA, usually having a *single* stranded *nucleotide chain* structure.

The monomeric units of nucleic acids are *nucleotides.*

2.1.1. Nucleotides

Each nucleotide is composed by a

- Phosphate group.
- 5 Carbon sugar (*ribose* in RNA, *deoxyribose* in DNA).
- Nitrogenous base that can be one of the following:

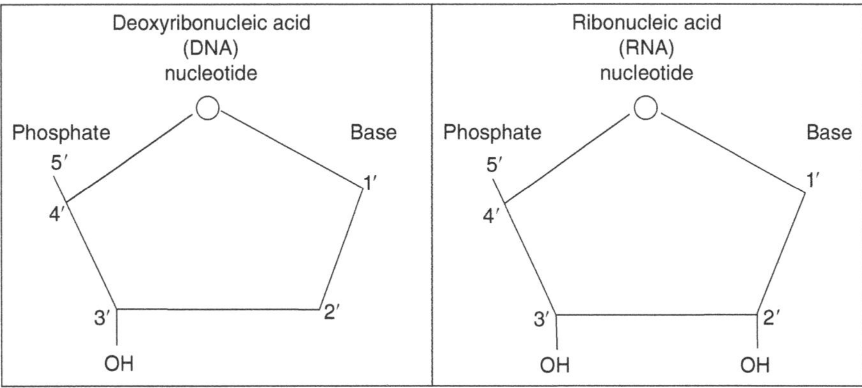

Fig. 1. Nucleotides' chemical structure.

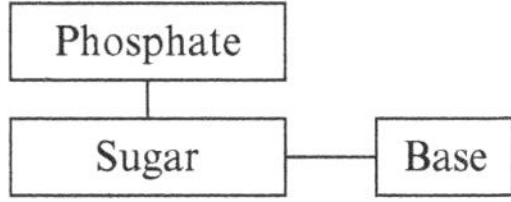

Fig. 2. Nucleotide schematic representation.

Purines
Adenine (A) in DNA and RNA
Guanine (G) in DNA and RNA
Pyrimidines
Cytosine (C) in DNA and RNA
Thymine (T) in DNA
Uracil (U) in RNA

The chemical structure of two nucleotides is shown in Fig. 1 where

- The 5 carbon sugar molecule is represented by a pentagon, the carbon positions are indicated by 1′, 2′, 3′, 4′, 5′.
- The nitrogenous base is held to the carbon in the sugar 1′ position.
- The phosphate group is joined to the 5′ sugar position.
- The nucleotide has a free hydroxyl group in the 3′ position (Fig. 2).

2.1.2. Polynucleotide Chain

Nucleotides join giving a polynucleotide chain (Fig. 3). For both DNA and RNA the union is between the 5′ phosphate group ($-PO_4$) of one of the nucleotides and the 3′ hydroxyl group (–OH) of the sugar of the other nucleotide by a phosphodiester bond.

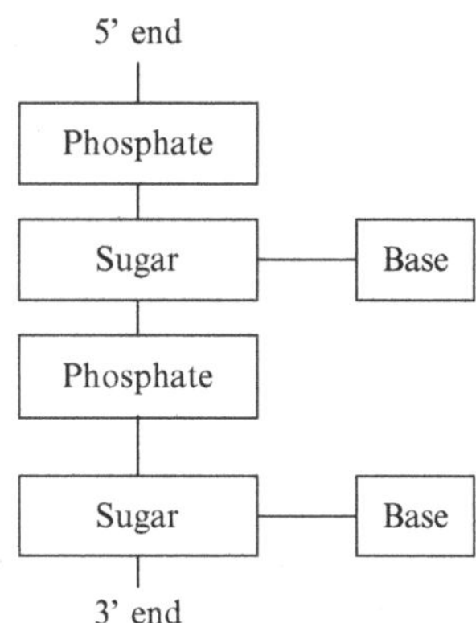

Fig. 3. A simple nucleotide chain (strand) of two bases.

One end of the nucleic acid polymers has a free hydroxyl (the 3′ end), the other end has a phosphate group (the 5′ end).

This directionality, in which one end of the DNA (or RNA) strand is chemically different than the other, is very important because DNA strands are always synthesized in the 5′ to 3′ direction. This has determinant implications in microarray experiments.

Any nucleotide chain is identified by its bases written in their sequential order. Sequences are always written from 5′ to 3′ ends. For example, a nucleotide chain of 6 nucleotides (and 6 bases) can be: ACGTTA.

2.1.3. Oligonucleotides

Oligonucleotides or *oligos* are short nucleotide chains of RNA or DNA. These sequences can have 20 or less bases (or pairs when they are double stranded).

50–70 nucleotide sequences are referred as *long* oligonucleotides, or simply *long oligos*, and play an important role in microarray technologies.

2.2. Structures

2.2.1. DNA Structure

The DNA structure consists of a polynucleotide *double chain* (or double strand) held together by weak bonds (hydrogen bonds) between the bases according to the following complementary base pairing rules

C≡G (with 3 hydrogen bonds).

A=T (with 2 hydrogen bonds).

in accordance with James Watson and Francis Crick 1953 model. The sequence of one of the strands determines the complementary sequence of the other strand.

Hydrogen bonds are weaker than the phosphodiester bonds in the alternating molecules of sugar and phosphate in DNA skeleton. These binding strength differences allow the separation of the two strands under special conditions while keeping the chain structure. Denaturation and hybridization processes that we will

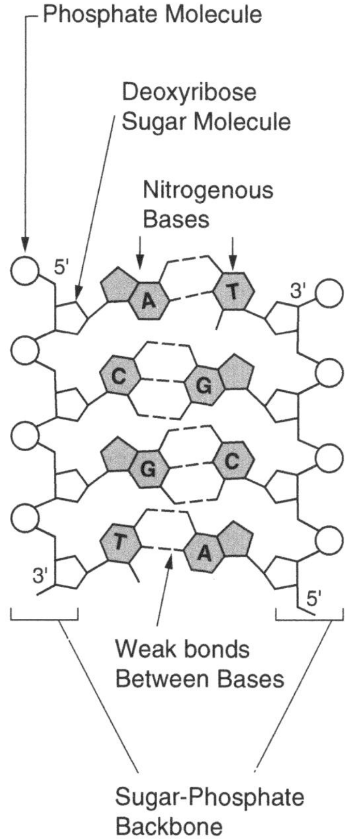

Fig. 4. DNA double stranded structure. Modified from http://genomics.energy.gov/gallery/basic_genomics/detail.np/detail-14.html.

see in Subheading 3 with relation to microarray experiments are deeply related to these hydrogen bonds.

Figure 4 shows a four bases DNA double chain. The hydrogen (weak) bonds between the bases are shown with broken lines and the double and triple bonds are explicitly differentiated. Also single ring pyrimidines (C, T, U) and double ring purines (A, G) can be appreciated as well as the 5′ to 3′ directions of the complementary strands.

Watson and Crick model also states that the two polynucleotide strands in the DNA molecule are wounded in a *double helix* as a twisted ladder with a sugar phosphate skeleton in the sides and nitrogen bases in the inside as rungs. Each DNA strand is half of the ladder.

In 1962 Francis Crick, James Watson, and Maurice Wilkins jointly received the Medicine Nobel prize for their 1953 DNA model based on Rosalind Franklin's work, as a molecular biologist and crystallographer. Rosalind who died of cancer in 1958 at the age of 37 could not receive the prize.

2.2.2. RNA Structure

As we have mentioned, the RNA is a *single stranded* polynucleotide and with the same bases as DNA except for the Thymine (T) that is replaced by Uracil (U) and the sugar is ribose instead of *deoxyribose* as in DNA.

2.3. A Eukaryotic Cell

Eukaryotic is a term that identifies a cell with a membrane-bounded nucleus in contrast with prokaryotic cells that lack a distinct nucleus (e.g., bacteria). A cell's genome is its total DNA content. Within the cell, besides the nuclear DNA of chromosomes, there are organelles in the cytoplasm called the mitochondrion with its own DNA. We will only consider nuclear DNA.

2.4. Human Genome

The nucleus of every human cell contains 46 chromosomes (23 pairs). Each chromosome basically consists of a long DNA double chain of approximately 2.5×10^7 nucleotides and base pairs. Unwounded, this chain can be up to 12 cm long. The human genome consists of approximately 3×10^9 base pairs.

Almost all of our cells have the same genetic information. What makes a liver cell different from a skin cell? The difference results from the fact that different genes are expressed at different levels. So

1. What is a gene?
2. What does it mean that a gene is expressed?

We will call gene a DNA sequence that contains the necessary information for the synthesis of a specific product.

The answer to the second question is in the following section.

2.5. The Central Dogma of Molecular Biology

The central dogma of molecular biology states that the information flows from DNA to RNA and then to protein. A portion of chromosomal DNA is copied (transcription process) into a single stranded messenger RNA (mRNA) that leaves the nucleus carrying the necessary information to synthesize a protein (translation process). Any sequence (or gene) that is active in this way is called "expressed." Although a reverse process from RNA to cDNA is possible (reverse transcription) the reverse process from protein to RNA has never been obtained.

2.5.1. Transcription

During the transcription process only one DNA strand is copied into a RNA. The synthesis of the single stranded RNA proceeds in its 5′ to 3′ direction. One strand of DNA directs the synthesis of the complementary mRNA strand. This DNA strand being transcribed is called the *template* or antisense strand. The other DNA strand is called the sense or coding strand. The RNA strand newly synthesized (primary gene transcript) contains the same information as the coding strand with the same base sequence with a U instead of a T and is complementary to the template strand.

Splicing

In the primary gene transcript or pre-mRNA there are segments that leave the nucleus after transcription and play an active role in the protein codification process, they are called exons. There are

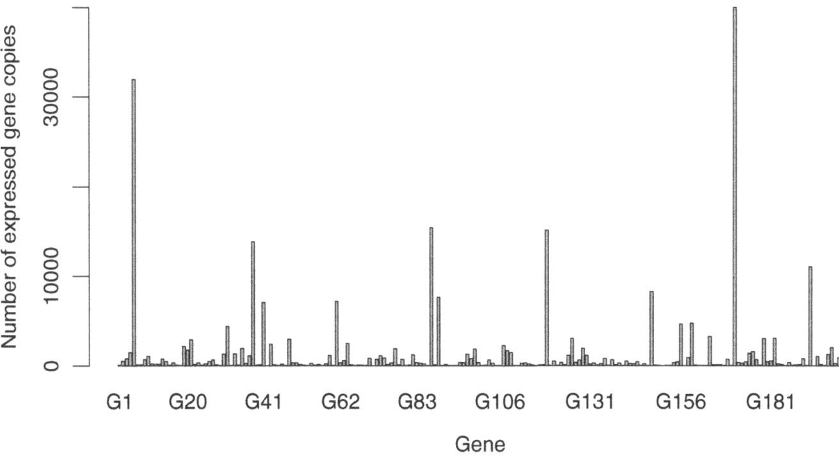

Fig. 5. Real hypothetic expression profile.

also segments called introns which are part of the transcribed mRNA that does not leave the nucleus. This RNA modification in which introns are removed and exons are joined is called splicing.

The messenger RNA that leaves the nucleus only has exons and in general is shorter than the original DNA template segment; it is the mature mRNA that has suffered the capping (G), polyadenylation (AAAA…) and splicing processes.

Alternative splicing is the RNA splicing variation mechanism in which the exons of the primary gene transcript, the pre-mRNA, are separated and reconnected to produce alternative ribonucleotide arrangements. Alternative splicing allows the synthesis of a greater variety of proteins than the originally DNA segments expressed have (2).

We will not describe the translation process that directs the mRNA; however, it is important to keep in mind that the amount synthesized is relatively proportional to the amount of mRNA transcribed. It is that amount of mRNA transcribed what we call *gene expression level.*

2.5.2. Gene Expression Profile

If we could count the number of mRNA molecules for each gene in a single cell we would obtain its "real expression profile." Figure 5 shows a "real hypothetic expression profile."

3. Microarray Technologies and Basic Principles They Rely On

In a microarray experiment the natural process determined by the central dogma of molecular biology is interrupted to extract mature mRNA from one or more tissues or cell lines to hybridize it (we'll soon see what this is) to its complementary cDNA previously fixed on the microarray. The microarray works as a detector of the amount and kind of mRNA present in the interrogated sample tissue.

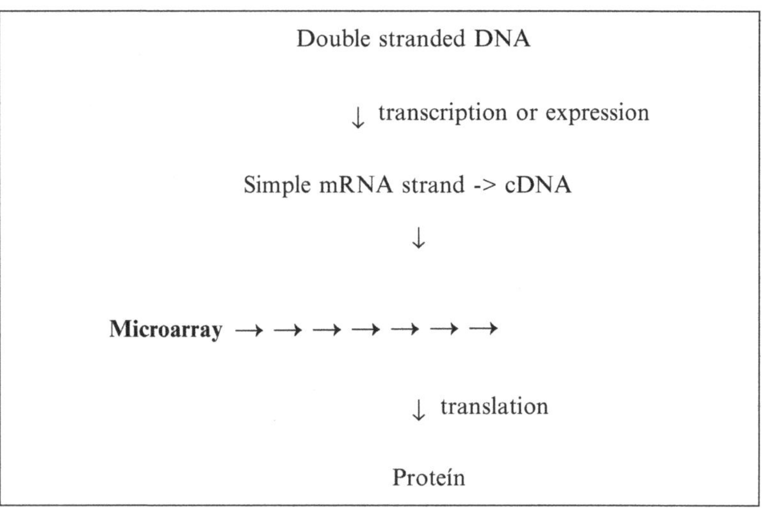

3.1. What Are DNA Microarrays?

DNA microarrays are small (2.5 cm × 6. 2.5 cm for spotted microarrays and 1.28 cm × 1.28 cm for high-density chips), solid supports onto which the thousands (10,000–1,000,000) of different DNA sequences are immobilized, or attached, at two dimensional fixed matrix locations called spots or features. Each spot contains millions of "identical" sequences.

- Each spot representing a different sequence has a unique physical location.
- May or may not have knowledge of the sequence.

The supports can be usual glass microscope slides, silicon chips, or nylon membranes.

According to different array manufacturing technologies (platforms),

- DNA is printed, spotted, or actually synthesized directly onto the support.
- Features or spots can either be approximately circles or rectangles.
- Fixed sequences—DNA, cDNA, short or long DNA oligonucleotides—are called probes.
- Microarrays allow one-color (channel) or two-color experiments.

3.2. Types of Microarrays

Microarrays can be classified according to the kind of the immobilized genomic sequences (probes). This is important as the probe sequences in the array identify complimentary sequences in the unknown sample genomic sequences (targets).

3.2.1. Transcriptomic Microarrays

The DNA immobilized to the array are complementary DNA sequences (cDNA) derived from known transcribed mRNA sequences or possible transcribed sequences (putative genes) for a certain type of tissue, with the purpose of measuring the amount of copies of the genes that are transcribed in a moment in the experimental tissue. These are called gene expression microarrays.

3.2.2. Comparative Genomic Hybridization Array

In Comparative Genomic Hybridization (CGH) arrays each spot contains DNA cloned sequences with known chromosomal location. This allows detecting gains and losses in chromosomes. Usually probes that map to evenly spaced loci along the entire length of the genome are printed. Also large pieces of genomic DNA can serve as the probed DNA.

3.2.3. Polymorphism Analysis Array

To detect mutations, immobilized DNA is usually from polymorphic variants of a single gene. The probed sequence placed on any given spot within the array will differ from that of other spots in the same microarray, sometimes by only one (Single Nucleotide Polymorphism, or SNP) or a few specific nucleotides.

3.3. Basic Principles on Which Microarray Experiments Rely On

DNA microarrays rely on the complementary rule: under adequate experimental conditions, complementary single stranded nucleic have strong tendency of binding in a double stranded nucleic acid molecule.

For every mRNA sequence of interest (*target*) a complementary DNA sequence (cDNA) can be obtained to immobilize a *probe* for that sequence onto the solid support. The position of the probe in the array identifies the sequence.

3.3.1. Nucleic Acid Hybridization and Denaturation

The chemical process by which two complementary single stranded nucleic acid chains zipper up to form a double stranded molecule is called *hybridization*.

When double stranded DNA molecules are subjected to conditions (pH, temperature, etc.) that disrupt their hydrogen bonds, the strands are no longer held together. This means that the strands separate as individual coils, it is then said that the double helix is *denatured*.

The denaturation conditions differ according to the relative G+C content in the DNA. The higher the G+C content of a DNA, the higher its denaturation temperature because G–C pairs are held by three H bonds whereas A–T pairs have only two.

The DNA denaturation is reversible. This process is called *DNA renaturation* or hybridization.

Hybridization reactions can occur between any (even those coming from different species) complementary single stranded nucleic acid chains: DNA/DNA, RNA/RNA, DNA/RNA.

Both denaturation and hybridization processes are important in microarray experiments.

3.4. Sample Genetic Material

3.4.1. mRNA Isolation

As we have already seen, the normal cellular modification of mRNA includes the polyadenylation process, this is the addition of up to 200 adenine nucleotides to one end of the molecule called poly(A) tail. In order to isolate mRNA from a given tissue, its cells are broken up and the cellular contents are exposed to beads coated with strings of thymine nucleotides. Because of adenine and thymine binding affinity the poly(A) mRNA is selectively retained on the beads while the other cellular components are washed away.

3.4.2. Reverse Transcription

Once isolated, purified mRNA is converted to single stranded DNA using the enzyme *reverse transcriptase* and is then made into a stable double stranded DNA using the enzyme *DNA polymerase*. DNA produced in this way is called *complementary DNA* (cDNA) because its sequence, at least the first strand, is complementary to that of the mRNA from which it was made, and represents only exon DNA sequences.

4. Microarray Experiments. General and Specific Remarks

In gene expression microarrays experiments the amount of mRNA (in a given tissue at a given moment for each sequence probed in the array) is measured indirectly using dye labelled molecules in order to answer, for example:

- How gene expression differs in different cell types.
- How gene expression differs in a normal and diseased (e.g., cancerous) cell.
- How gene expression changes when a cell is treated by a drug.
- How gene expression changes when the organism develops and cells are differentiating.
- How gene expression is regulated—which genes regulate which and how.

There are six general steps to follow in the microarray experiments:

- Relevant questions, statistical experiment design.
- Microarray manufacturing.
- Sample preparation and target labelling.
- Hybridization of the labelled target samples sequences to the corresponding microarray probes.
- Washing to eliminate the excess solution and reduce the nonspecific binding.
- Scanning the microarray under laser light and obtaining a digital image.

The initial data resulting from microarray experiments are one or two digital images, depending on the microarray platform used, for every microarray. Three more data analysis steps follow:

- Image analysis.
- Calibration.
- Statistical data analysis.

4.1. Design

Many experimental researchers, believe that statistical issues can be of secondary importance at the early stages of the experiment and that statisticians should be incorporated at the data analysis and interpretation phase of the investigation. However, also in this research area, data analysis cannot compensate for inadequate design. For microarrays experiments it is important to remember that different experimental conditions may give different expression profiles for the same biological setting (3).

The proposals for experimental design- and model-based analysis taking into account random variability are not new (4–6). We will only describe a few specific aspects regarding the array design and the samples design.

4.1.1. Array Design

The choice of the DNA probe sequences to be synthesised or spotted on to the array depends on the technology and the type of genes the researcher wishes to interrogate or by the cDNA libraries (collections of cDNA clones) available. For high-density in situ synthesized short oligos (25 bases) microarrays it is mainly the manufactures' decision; however, specific custom arrays can also be ordered at higher costs. Many researchers also buy or make their own spotted cDNA or long oligos (60–70 bases) microarrays. In the design of these arrays they must decide:

- Which and where the probes will be spotted.
- Which and where the controls will be spotted.

Controls are special probes included in the array, some have the purpose of evaluating the quality of the experiment and others will be used to standardise the measurements. The usual control probes are:

- *Negative controls*: Empty spots, buffer solution spots.
- *Level controls*: Spots with cDNA or oligos from different species that will not interfere with the sampled genomic sequences (i.e., bacterial DNA if mammals are studied) complementary sequences will be added to the samples (*spiked in*) in pre specified quantities. Their intensity values serve for calibration.
- *Positive controls*: "Housekeeping genes," highly expressed genes in all samples to evaluate if the hybridization has effectively occurred.

Adjacent d*uplicate spots* (two or more times) are included to evaluate signal variation. However this estimation will be lower than the real meaningful between array variability for a given spot among replicated arrays.

4.1.2. Sample Design

Technical Replicates

When multiple microarrays are hybridized with mRNA from a single biological case we have technical replicates. These replications only allow measuring the variability due to measurement errors and are useful in quality-control studies.

Biological Replicates

Biological replicates allow the evaluation of both measurement variability and biological differences between cases. This type of replication is obtained when the sample mRNA comes from different individuals from a given species or different cell lines and is required when the aim is to make inferences about populations. Although early microarray experiments used few or no biological replicates, their necessity is now undisputed (7).

Sample Size

Early microarray studies (8–10) used a single "two channel" microarray or two "one channel" microarrays to identify differentially expressed genes. Several proposals were developed to deal with the problem (11–13). The idea behind the proposals was that the measurements on many genes, i.e., variables, could compensate the reduced sample size. Sizes of 2 or 3 were considered large. This point of view is changing to realize that even if technical variability was eliminated it is not possible to reduce random variability inherent to biological processes and that there is no alternative to increasing samples sizes in microarray studies (14).

Observational Studies

In many microarray experiments sample DNA or RNA come from observational studies. Bias and confounding factors should have been considered as they are in any epidemiological study; however this has not been the practise in such experiments (15, 16) which in general lack of standard epidemiological approaches (i.e., assessment of chance, bias, and confounding). The advantages that microarray technology can introduce in clinical and epidemiological studies, if well established epidemiologic principles are not sacrificed in the process, is beginning to be noticed (17).

5. Image Analysis

The resulting data from a microarray experiment are one or two digital images for each microarray in the experiment. These microarray digital images (18) provide a snapshot of the types and quantities of molecules that have reacted during hybridization and hence were present in the sample targets.

A microarray image is a two-dimensional numeric representation in which each value gives the mean intensity of a small sector or pixel. Every spot is represented by hundreds of pixels and we have *pixel wise data*. It is necessary to obtain a summary intensity measure for each spot in the array.

Image processing can be divided in three tasks

- *Gridding*: It is the assignment of spots' coordinates.
- *Segmentation*: Classification of foreground (signal) and background pixels.
- *Target intensity extraction*: Obtaining a summary measure of the spot intensity from the foreground and background pixels.

Perhaps the most critical aspect of image processing is segmentation; this is deciding which pixels correspond to the signal for each spot. There are four commonly used segmentation methods: fixed circle, variable circle, histogram, and adaptive shape. It has not yet been established and it is not clear if there exists an optimal method (19–23). High-density oligonucleotide spots are square, and squared regions are considered for the spot foreground and background summary measures (24).

Standard image processing methods subtract background from foreground intensities to obtain the final intensity value for each pixel. This background correction gives negative signals for microarray images. Several proposals (19–22) deal with this drawback, as missing values are artificially generated with the usual base 2 logarithmic ($\log_2$) data transformation used. Image analysis including background correction methods still is an active research area.

Probe wise data are the final result of the image processing stage. These are the data we are considering in subsequent sections.

Data from high-density microarrays require a preliminary summarizing step, that of probe intensities that interrogate the same genomic sequence (25, 26).

6. Data Calibration

Data from microarray experiments show two types of problems that are faced through data transformations:

1. *MA* plots present curved structures not attributable to biological reasons.
2. Probe intensity variability is mean dependent.

Let Y_{rk} represent the intensity of probe k in array r that resulted from the image processing stage. *MA* plots compare the intensity of two "one channel" arrays (or the two channels of a two channel microarray) in scatter plots of

$M_k = \log_2 (\Upsilon_{1k} / \Upsilon_{2k})$ in the *y*-axes.
$A_k = 0.5 \log_2 (\Upsilon_{1k} \times \Upsilon_{2k})$ in the *x*-axes.

If the samples of both arrays (or channels of the same array) come from identical biological conditions (self–self experiment) no tendency is expected in a *MA* plot. However curved structures not attributable to biological reasons are usually seen in these plots.

A very frequently used procedure that eliminates such structures consists in subtracting, to every *M* value in the plot, the fit of a local smoother in the original *MA* plot. The data on that curve are representing the probes with no different expression between the compared arrays (27). The transformed data can be written as

$$Z_{rk} = \log_2(\Upsilon_{rk}) + \frac{C_k}{2}, \tag{1}$$

where C_k is a constant depending on the spot and the local smoother. This procedure is straight forward and flexible enough to capture most of the structures appearing in *MA* plots but we are forcing the data to satisfy our expectations.

The data transformation given in Eq. 1 solves the first of the two problems stated at the beginning of this section and partially the second one.

In relation to the mean variance dependence of microarray intensity data several authors coincide in modelling intensities through a multiplicative additive model (28, 29)

$$\Upsilon_{rk} = a_r + b_r X_{rk} e^{\eta_k + \varsigma_{rk}} + \varepsilon_k + \delta_{rk}, \tag{2}$$

where X_{rk} is the true intensity of spot *k* in array *r*; a_r, b_r are constants and $\eta_k, \varsigma_{rk}, \varepsilon_k, \delta_{rk}$ are error terms. Moreover the following transformation

$$Z_{rk} = \log(B_r \Upsilon_{rk} + C_r + \sqrt{(B_r \Upsilon_{rk} + C_r)^2 + 1}),$$

has independently been proposed by several authors (30–32) to stabilize the variance in microarray data that satisfy the multiplicative additive model (2) and the array dependent constants B_r, C_r are estimated from the data. This transformation is based on a quadratic relationship between the variance and signal strength in the original scale if the data meet the model (2). This model can also explain all structures that are found in *MA* plots (33). Even if only the nonrandom components are considered, the nonlinear dependencies can be explained and removed, giving a much simpler interpretation: the true intensities are related to the observed intensities through an affine transformation

$$\Upsilon = a + b \cdot X.$$

and different choices for the parameters of the affine transformation for different arrays give the patterns observed in *MA* plots (34).

7. Statistical Analysis

7.1. Inference

One of the basic statistical inference concerns is to conduct tests to decide whether the difference of two sample means provides enough evidence to decide that the population means are different. In a microarray context it comes to detect genes with mean different expression levels between two or more groups (types of tissue) that provide enough evidence to decide that the genes are differentially expressed. The problem arises from the fact that thousands of tests must be conducted, one for each genomic sequence of interest.

There is a widespread agreement that multiple comparison procedures should be used. The usually recommended procedure is the false discovery rate (FDR). The argument in favour of using this error rate is based on the fact that the usual correction which is obtained by using the classical Bonferroni procedure is too restrictive resulting in a reduction in power. However the strong coregulation gene structures lead to very unstable estimation of the FDR. A Bonferroni type procedure controlling the expected value of false positives results in more stable estimates than those from FDR in comparable powers (35).

Multiple comparisons can be reduced and power improved through the comparison of a priori defined subsets of genes; the subsets are tested between two biological states in what is called "gene subset enrichment analysis (GSEA)" (36). This 2003 proposal, that uses groups of genes that share common biological function, chromosomal location, or regulation, has been used in a number of applications (37–39) and is going through subsequent improvements (40–47).

7.2. Classification

Unsupervised classification is one of the first statistical techniques used in the analysis of microarray data and is one of the favourites. This method attempts to divide the data into classes without prior information (unsupervised classification) or predefined classes. It has shown some successes in finding relevant and meaningful patterns (48–51). However, the researcher is guaranteed to obtain gene clusters, regardless of

- Sample size.
- Data quality.
- The design of the experiment.
- Any other biological validity that is affiliated with the grouping.

Unsupervised classification should be avoided, if it is inevitable, some sort of reproducibility measure should be provided. Those procedures that re-sample at case level—rather than gene level—have a reasonable performance and none is considered the best.

Supervised classification procedures need an independent cross validation as the resulting prediction rules are based on a relatively small number of samples of various types of tissues containing expression data of many thousands of genes.

The results of classification procedures may be representing to the data too much giving low or null predictive power; this is what is called over fitting.

8. Challenges

Microarray technologies generated explosive expectations related to the advances that in biology and medicine would occur in the short term (51, 52) but as results did not parallel those expectations the literature reflected the disappointments (53). As the technology of microarrays ran ahead of analysis techniques, researchers from various fields carried out their own statistics to analyze the data (54).

A growing number of publications per year from the year 1995 appeared since Schena (55) presented his first work on such experiments. Figure 6 shows the number of microarrays publications per year (selecting in Pub Med for the keywords "microarray or microarrays"). The number of articles per year has an exponential growth from 1995 to 2001. From then until 2005 the growth is linear with about a thousand more publications every year. The value of 2006 deviates slightly below this trend and in 2007 the deviation is even further.

8.1. Some Successes

In the decade since the beginning of technology, there have also been successes. Patients with leukaemia were automatically and accurately classified in two main subtypes of the disease using only

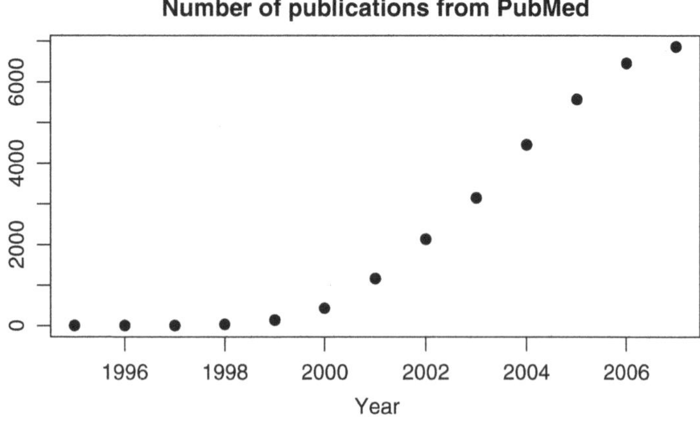

Fig. 6. Number of publications selected with "microarray or microarrays" keywords.

gene expression levels in 1999 (56). While these forms of leukaemia were already known and well characterized, the experiment showed that the strategy could in principle reveal unknown subtypes. Further in 2001, researchers identified five patterns of gene expression levels in breast cancer (57) and showed that corresponded to different types of diseases with different prognosis.

More recently in 2006 a gene with a fold-change of 50 times the level of expression in cancer patients who did not respond to chemotherapy treatment in comparison to those who did respond to treatment was found (58). This gene encodes for a protein that prevents tumour cell death; blocking this protein might allow a chemotherapy response.

In 2005, the US Food and Drug Administration (FDA) approved the first microarray-based clinical test. The test identifies genetic variations in two key codifying regions CYP2D6 and CYP2C19 for the cytochrome P450 enzyme that metabolizes usual drugs. This will enable doctors to personalize drug choice and dosing (59).

8.2. Frustrations

Also, in the decade since the advent of microarray technology a great deal of frustration has accumulated among biologists who have dedicated their efforts in following up false research directions. Many articles have been discredited; scientists have difficulties in finding studies that point to something concrete and in validating the results and several articles show this frustration (60–62). Several studies (63–65) have found that the list of differentially expressed genes has had very low overlap between different platforms. However the overlap among independent studies for the same biological question can have important improvements if coherent statistical analysis are carried out (66).

Numerous discussions in the literature show a tendency to explain the glaring lack of power and instability of the results of data analysis, by a high technical level of noise in the data.

The Quality Control (MAQC) Consortium project has generated public available databases that may give an answer to the referred discussions (67). These technical replicates addresses the evaluation of

- Repeatability within a single site.
- Reproducibility between sites.
- Comparability between platforms.

Several papers reporting the analysis of MACQ data (68–73) are showing promising results on the reliability and reproducibility of microarray technology and reflecting that the random fluctuations of gene expression signals caused by technical noise are quite low.

So, how are problems explained? Fig. 7 shows the number of publications selected with "microarray or microarrays" and "statistics or statistical" keywords. Its striking the low number of microarray publications presenting statistical analysis.

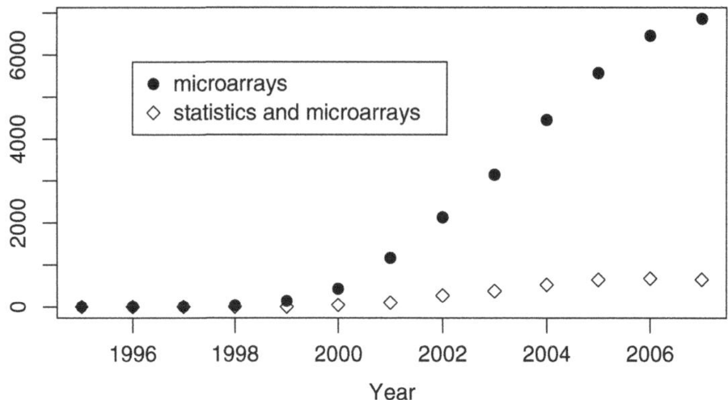

Fig. 7. Number of publications selected with "microarray or microarrays" and "statistics or statistical" keywords.

9. Conclusions

Microarray experiments present two main types of difficulties. The first one is microarray reliability. The MACQ project currently addresses the evaluation of the repeatability within a single site, between sites reproducibility and comparability between platforms. Although some researchers believe that experiments addressed in the MAQC project are conducted in conditions too "ideal" and that hardly reflect the real situation of many experimental laboratories (74) and sample sizes are not large enough (75), in general the results are satisfactory (76). Moreover, this technology is constantly evolving and improving, so it is expected to provide better and lower budget results.

The second difficulty lies in the design of the experiment and data analysis. It is important to incorporate epidemiological principles to the experiment design whenever dealing with observational studies. Also, microarray research area would benefit from identical design and statistical analysis for different experiments on a given biological problem. The second aspect of this goal would be achieved if researchers made available their raw data for reanalysis. Even better would be if research on one topic conducted by independent labs would follow the same protocol. This, together with the reduction of the experimental costs, may increase sample sizes and thus an improve power. A deeper understanding of normal biological variability and gene coregulation mechanisms could be addressed (77, 78). It is critical to the advancement of knowledge in molecular biology that microarrays no longer be simply used as exploratory tools.

References

1. http://www-lbit.iro.umontreal.ca/ISMB98/anglais/ontology.html
2. Lopez AJ (1998) Alternative splicing of pre-mRNA: developmental consequences and mechanisms of regulation. Annu Rev Genet 32:279–305
3. http://www.affymetrix.com/support/technical/technotes/blood_technote.pdf
4. Churchill GA (2002) Fundamentals of experimental design for cDNA microarrays. Nat Genet 32:490–495
5. Kerr MK, Churchill GA (2001) Statistical design and the analysis of gene expression microarray data. Genet Res 77(2):123–128
6. Smyth GK, Yang YH, Speed T (2003) Statistical issues in cDNA microarray data analysis. Methods Mol Biol 224:111–136
7. Allison D, Cui X, Page G, Sabripour M (2006) Microarray data analysis: from disarray to consolidation and consensus. Nat Rev Genet 7:55–65
8. DeRisi J, Penland L, Brown PO, Bittner ML, Meltzer PS, Ray M, Chen Y, Yan AS, Trent JM (1996) Use of a cDNA microarray to analyse gene expression patterns in human cancer. Nat Genet 14:457–460
9. Schena M, Shalon D, Davis RW, Brown PO (1995) Quantitative monitoring of gene expression patterns with a complementary DNA microarray. Science 270:467–470
10. Schena M (1996) Genome analysis with gene expression microarrays. BioEssays 18:427–431
11. Chen Y, Dougherty E, Bittner M (1997) Ratio-based decisions and the quantitative analysis of cDNA microarray images. J Biomed Opt 2(4):364–374
12. Newton M, Kendziorskim M, Richmond C, Blattner F, Tsui K (2001) On differential variability of expression ratios: improving statistical inference about gene expression changes from microarray data. J Comput Biol 8(1):37–52
13. Sapir M, Churchill GA (2000) Estimating the posterior probability of differential gene,expression from microarray data. Poster, The Jackson Laboratory. http://www.jax.org/research/churchill/pubs/marina.pdf
14. Klebanov L, Yakovlev A (2007) Is there an alternative to increasing the sample size in microarray studies? Bioinformation 1(10):429–431
15. Potter JD (2001) At the interfaces of epidemiology, genetics, and genomics. Nat Rev Genet 2:142–147
16. Potter JD (2003) Epidemiology, cancer genetics and microarrays: making correct inferences, using appropriate designs. Trends Genet 19(12):690–695
17. Webb PM, Melissa A, Merritt MA, Boyle MG, Green AC (2007) Microarrays and epidemiology: not the beginning of the end but the end of the beginning. Cancer Epidemiol Biomarkers Prev 16:637–638
18. Schena M (2003) Microarray analysis. Wiley-Liss, Hoboken, NJ. ISBN 9780471414438
19. Yang YH, Buckley MJ, Speed TP (2001) Analysis of cDNA microarray images. Bioinformatics 2(4):341–349
20. Yang YH, Dudoit S, Luu P, Lin DM, Peng V, Ngai J, Speed TP (2002) Normalization for cDNA microarray data: a robust composite method addressing single and multiple slide systematic variation. Nucleic Acids Res 30:e15
21. Angulo J, Serra J (2003) Automatic analysis of DNA microarray images using mathematical morphology. Bioinformatics 19(5):553–562
22. Li Q, Fraley C, Bumgarner R, Yeung K, Raftery A (2005) Donuts, scratches and blanks: robust model-based segmentation of microarray images. Technical Report no. 473. Department of Statistics, University of Washington
23. Ahmed A, Vias M, Iyer N, Caldas C, Brenton J (2004) Microarray segmentation methods significantly influence data precision. Nucleic Acids Res 32(5):1–7
24. Wu Z, Irizarry R, Gentleman R, Murillo F, Spencer F (2003) A model based background adjustment for oligonucleotide expression arrays CGRMA-MLE. Technical Report, John Hopkins University, Department of Biostatistics, Baltimore, MD. Working Papers
25. Irizarry R, Hobbs F, Beaxer-Barclay Y, Antonellis K, Scherf U, Speed T (2003) Exploration, normalization, and summaries of high density oligonucleotide array probe level data. Biostatistics 4:249–264
26. Irizarry RA, Bolstad BM, Collin F, Cope LM, Hobbs B, Speed TP (2003) Summaries of Affymetrix GeneChip probe level data. Nucleic Acids Res 31(4):e15
27. Yang YH, Dudoit S, Luu P, Lin DM, Peng V, Ngai J, Speed TP (2002) Normalization for cDNA microarray data: a robust composite method addressing single and multiple slide systematic variation. Nucleic Acids Res 30:e15
28. Durbin BP, Hardin JS, Hawkins DM, Rocke DM (2002) A variance estabilizing transformation for gene expression microarray data. Bioinformatics 18:105–110

29. Huber W, Von Heydebreck A, Sultmann H, Poustka A, Vingron M (2002) Variance stabilization applied to microarray data calibration and to the quantification of differential expression. Bioinformatics 18:96–104
30. Munson P (2001) A "consistency" test for determining the significance of gene expression changes on replicate samples and two-convenient variance-stabilizing transformations. *GeneLogic Workshop* on Low Level Analysis of Affymetrix GeneChip Data, Nov. 19, Bethesda, MD
31. Durbin BP, Hardin JS, Hawkins DM, Rocke DM (2002) A variance estabilizing transformation for gene expression microarray data. Bioinformatics 18:105–110
32. Huber W, von Heydebreck A, Sueltmann H, Poustka A, Vingron M (2003) Parameter estimation for the calibration and variance stabilization of microarray data. Stat Appl Genet Mol Biol 2:3.1–3.22
33. Cui X, Kerr M, Churchill G (2003) Transformations for cDNA microarray data. Stat Appl Genet Mol Biol 2(1) Article 4
34. Bengtsson H, Hössjer O (2006) Methodological study of affine transformations of gene expression data with proposed robust non-parametric multi-dimensional normalization method. BMC Bioinform 7(100):1–18
35. Gordon A, Glazko G, Qiu X, Yakovlev A (2007) Control of the mean number of false discoveries, Bonferroni and stability of multiple testing. Ann Appl Stat 1(1):179–190
36. Mootha VK, Lindgren CM, Eriksson KF, Subramanian A, Sihag S, Lehar J, Puigserver P, Carlsson E, Ridderstrale M, Laurila E, Houstis N, Daly MJ, Patterson N, Mesirov JP, Golub TR, Tamayo P, Spiegelman BM, Lander ES, Hirschhorn JN, Altshuler D, Groop LC (2003) PGC-1alpha-responsive genes involved in oxidative phosphorylation are coordinately downregulated in human diabetes. Nat Genet 34:267–273
37. Lamb J, Ramaswamy S, Ford HL, Contreras B, Martinez RV, Kittrell FS, Zahnow CA, Patterson N, Golub TR, Ewen ME (2003) A mechanism of cyclin D1 action encoded in the patterns of gene expression in human cancer. Cell 114(3):323–334
38. Majumder PK, Febbo PG, Bikoff R, Berger R, Xue Q, McMahon LM, Manola J, Brugarolas J, McDonnell TJ, Golub TR, Loda M, Lane HA, Sellers WR (2004) mTOR inhibition reverses Akt-dependent prostate intraepithelial neoplasia through regulation of apoptotic and HIF-1-dependent pathways. Nat Med 10(6):594–601
39. Isakoff MS, Sansam CG, Tamayo P, Subramanian A, Evans JA, Fillmore CM, Wang X, Biegel JA, Pomeroy SL, Mesirov JP, Roberts CW (2005) Inactivation of the Snf5 tumor suppressor stimulates cell cycle progression and cooperates with p53 loss in oncogenic transformation. Proc Nat Acad Sci U S A 102(49):17745–17750
40. Xiao Y, Frisina R, Gordon A, Klebanov LB, Yakovlev AY (2004) Multivariate search for differentially expressed gene combinations. BMC Bioinform 5(1):164
41. Dettling M, Gabrielson E, Parmigiani G (2005) Searching for differentially expressed gene combinations. Genome Biol 6:R88
42. Subramanian A, Tamayo P, Mootha VK, Mukherjee S, Ebert BL, Gillette MA, Paulovich A, Pomeroy SL, Golub TR, Lander ES, Mesirov JP (2005) Gene set enrichment analysis: a knowledge-based approach for interpreting genome-wide expression profiles. Proc Natl Acad Sci U S A 102(43):15545–15550
43. Tian L, Greenberg SA, Kong SW, Altschuler J, Kohane IS, Park PJ (2005) Discovering statistically significant pathways in expression profiling studies. Proc Natl Acad Sci U S A 102(38):13544–13549
44. Barry WT, Nobel AB, Wright FA (2005) Significance analysis of functional categories in gene expression studies: a structured permutation approach. Bioinformatics 1(9):1943–1949
45. Efron B, Tibshirani R (2007) On testing the significance of sets of genes. Ann Appl Stat 1(1):107–129
46. Klebanov L, Glazko G, Salzman P, Yakovlev A (2007) A multivariate extension of the gene set enrichment analysis. J Bioinform Comput Biol 5(5):1139–1153
47. Eisen MB, Spellman PT, Brown PO, Botstein D (1998) Cluster analysis and display of genomewide expression patterns. Proc Natl Acad Sci U S A 95(25):14863–14868
48. Golub TR, Slonim DK, Tamayo P, Huard C, Gassenbeek M, Mesirov JP, Coller H, Loh ML, Downing JR, Caligiuri MA, Bloomfield DD, Lander ES (1999) Molecular classification of cancer: class discovery and class prediction by gene expression monitoring. Science 286(15):531–537
49. Tamayo P, Slonim D, Mesirov J, Zhu Q, Kitareewan S, Dmitrovsky E, Lander ES (1999) Interpreting patterns of gene expression with self-organizing maps: methods and application to hematopoietic differentiation. Proc Natl Acad Sci U S A 96(6):2907–2912
50. Wen X, Fuhrman S, Michaelis GS, Carri DB, Smith S, Barker SJ, Somogyi R (1998) Large-scale temporal gene expression mapping of central nervous system development. Proc Natl Acad Sci U S A 95:334–339
51. Lander E (1999) Array of hope. Nat Genet (Supplement 21)

52. Schena M (2003) Microarray analysis preface page XIV. Wiley-Liss, Hoboken, NJ. ISBN 9780471414438
53. Frantz S (2005) An array of problems. Nat Rev Drug Discov 4:362–363
54. Cobb K (2006) Re inventing statistics in microarrays: the search for meaning in a vast sea of data. Biomed Comput Rev 2(4):21
55. Schena M, Shalon D, Davis RW, Brown PO (1995) Quantitative monitoring of gene expression patterns with a complementary DNA microarray. Science 270:467–470
56. Golub TR, Slonim DK, Tamayo P, Huard C, Caasenbeek M, Mesirov JP, Coller H, Loh ML, Downing JR, Caligiuri MA, Bloomfield CD, Lander ES (1999) Molecular classification of cancer: class discovery and class prediction by gene expression monitoring. Science 286:531–537
57. Sorlie et al (2001) Gene expression patterns of breast carcinomas distinguish tumor subclass with clinical implications. Proc Natl Acad Sci USA 98(19):10869–10874
58. Petty RD, Kerr KM, Murray GI, Nicolson MC, Rooney PH, Bissett D, Collie-Duguid ES (2006) Tumour transcriptome reveals the predictive and prognostic impact of lysosomal protease inhibitors in non-small-cell lung cancer. J Clin Oncol 24(11):1729–1744
59. http://www.medicalnewstoday.com/articles/18822.php
60. Frantz S (2005) An array of problems. Nat Rev Drug Discov 4:362–363
61. Ioannidis JPA (2005) Microarrays and molecular research: noise discovery? The Lancet 365(9458):454–455
62. Marshall E (2004) Getting the noise out of gene arrays. Science 306:630–631
63. Tan PK et al (2003) Evaluation of gene expression measurements from commercial microarray platforms. Nucleic Acids Res 31:5676–5684
64. Miller RM et al (2004) Dysregulation of gene expression in the 1-methyl-4-phenyl-1,2,3,6-tetrahydropyridine-lesioned mouse substantia nigra. J Neurosci 24(34):7445
65. Miklos GL, Maleszka R (2004) Microarray reality checks in the context of a complex disease. Nat Biotechnol 22:615–621
66. Suárez-Fariñas M, Noggle S, Heke M, Hemmati-Brivanlou, Magnasco M (2005) Comparing independent microarray studies: the case of human embryonic stem cells. BMC Genomics 6(99):1–11
67. MAQC Consortium (2006) The MicroArray Quality Control (MAQC) project shows inter- and intraplatform reproducibility of gene expression measurements. Nat Biotechnol 24(9):1151–1161
68. Bosotti R et al (2007) Cross platform microarray analysis for robust identification of differentially expressed genes. BMC Bioinform 8(Suppl 1):S5
69. Wang Y et al (2006) Large scale real-time PCR validation on gene expression measurements from two commercial long-oligonucleotide microarrays. BMC Genomics 7:59
70. Kuo WP et al (2006) A sequence-oriented comparison of gene expression measurements across different hybridization-based technologies. Nat Biotechnol 24(7):832
71. Canales RD et al (2007) Evaluation of DNA microarray results with quantitative gene expression platforms. Nat Biotechnol 24(9): 1115
72. Klebanov L, Yakovlev A (2007) How high is the level of technical noise in microarray data? Biol Direct 2:9
73. Robinson MD, Speed TP (2007) A comparison of Affymetrix gene expression arrays. BMC Bioinform 15(8):449
74. Perkel J (2006) Six things you won't find in the MAQC. Scientist 20(11):68
75. Klebanov L, Qiu X, Welle S, Yakovlev A (2007) Statistical methods and microarray data. Nat Biotechnol 25:25–26
76. http://www.microarrays.ca/MAQC_Review_July2007.pdf
77. Klebanov L, Jordan C, Yakovlev A (2006) A new type of stochastic dependence revealed in gene expression data. Stat Appl Genet Mol Biol 5:1
78. Klebanov L, Yakovlev A (2007) Diverse correlation structures in gene expression data and their utility in improving statistical inference. Ann Appl Stat 1(2):538–559

Chapter 3

Multiple Hypothesis Testing: A Methodological Overview

Anthony Almudevar

Abstract

The process of screening for differentially expressed genes using microarray samples can usually be reduced to a large set of statistical hypothesis tests. In this situation, statistical issues arise which are not encountered in a single hypothesis test, related to the need to identify the specific hypotheses to be rejected, and to report an associated error. As in any complex testing problem, it is rarely the case that a single method is always to be preferred, leaving the analysts with the problem of selecting the most appropriate method for the particular task at hand. In this chapter, an introduction to current multiple testing methodology was presented, with the objective of clarifying the methodological issues involved, and hopefully providing the reader with some basis with which to compare and select methods.

Key words: Multiple hypothesis testing, Gene expression profiles, Stepwise procedures, Bayesian inference

1. Introduction

Perhaps the central problem in the analysis of microarray data is the screening for genes which differentially express between two or more experimental conditions, thereby associating the role played by those genes with an experimental phenotype. A sample of gene expression profiles is collected from each condition, within which expressions are summarized, and a decision rule applied with which to decide whether or not a gene is differentially expressed. Often, conventional ad hoc rules are used (say, a 2.5-*fold change*), or genes are simply ranked according to the magnitude, or the statistical significance, of the observed differential expressions. Since such experiments are often exploratory, the decision of how many genes to tentatively regard as differentially expressed may depend on the resources available for experimental post hoc validation, rather than statistical analysis.

Andrei Y. Yakovlev et al. (eds.), *Statistical Methods for Microarray Data Analysis: Methods and Protocols*, Methods in Molecular Biology, vol. 972, DOI 10.1007/978-1-60327-337-4_3,

Whatever external considerations apply, it is important to note that statistical methodology is currently able to add quantitative precision to the gene discovery process. Here we will focus on two specific problems: (1) estimating the proportion of all screened genes which are differently expressed, and (2) constructing a list of tentative differentially expressed genes coupled with a reportable error. The two problems are related but distinct, and may assume different priorities depending on the application. For example, finding genes which differentially express in blood tissue for subjects with early Parkinson's disease, which normally affects brain tissue, has important implications for the development of clinically useful biomarkers (see ref. (1)), so that interest may be in the prevalence of differentially expressed genes rather than in their precise identification. While the second problem is usually of greater interest, some recent methods, discussed below, make use of differential expression prevalence in reporting gene selection errors.

These problems fall under the theory of multiple hypothesis testing, which has been a formal branch of statistical theory since John Tukey's pioneering work in the early 1950s (see ref. (2)). It is easy to see that if a sequence of hypothesis tests are performed with using a 5% level of significance, then an average of 1 out of 20 among true null hypotheses will be falsely rejected. In particular, in screening applications involving thousands of genes, we should anticipate that the false positives may greatly outnumber the true positives. We then need to make use of the magnitude of the observed significance level (p-value), recognizing that if we lower the significance level defining rejection of the null hypothesis, the proportion of false positives among those reported decreases also. The trade off, hence the need for analytical methods, lies in the fact that this also results in a decrease in the detection of true positives.

This branch of statistical methodology has fortunately achieved a level of conceptual unity, so that a large class of multiple testing procedures (MTP) can be described by a compact system of general definitions and structures. We start with a set of m null hypotheses H_1, …, H_m. The true state of affairs is described by unobserved parameters τ_i which are set to zero if H_i is true and one otherwise, for each $i=1$, …, m. The set of all true null hypotheses will be $T=\{i : \tau_i=0\}$. Associated with each H_i is a sample of data W_i which can be transformed to a p-value p_i. The crucial fact is that p_i is uniformly distributed if $\tau_i=0$, and is stochastically smaller than uniform otherwise. They may or may not be independent, and the validity of the null uniform distribution assumption must be considered. From these elements, the formal objective of the MTP is stated in Subheading 2.

One way of characterizing MTPs is in the relative importance placed on p_i or W_i. Perhaps the most widely used MTP is the *Bonferonni Procedure* (see below), which is implemented by selecting a p-value threshold α, and rejecting all hypotheses with p-values at or below that value. No reference to W_i is made once the p-values are calculated. This method is easy to implement and broadly applicable, but this comes at a cost of a sometimes pronounced conservatism, in the sense that the reported error is usually overestimated. In Subheading 3 we discuss a number of subsequent refinements of the Bonferonni procedure. Some of these refinements are purely structural, while others improve the error estimate by imposing conditions on the data (W_1, ..., W_m). Overall, since they are based on the selection of a p-value threshold, they retain some of the simplicity and broad applicability of the Bonferonni procedure.

A second class of MTPs is based on a type of mixture model driven by the hypotheses H_i. Let $T_i = T_i(W_i)$ be some statistic with marginal density $f(t) = \pi_0 f_0(t) + \pi_1 f_1(t)$. The values $(\tau_1,\ldots,\tau_m)$ are replications of a mixture parameter τ, so that f_0 and f_1 are the densities of T_i conditional $\tau_i = 0$ or $\tau_i = 1$ respectively. In addition, we have $E(\tau) = \pi_1$. Possibly, T_i is a p-value, in which case we may set f_0 to be the uniform distribution. We note, however, that the uniform null assumption (or other natural null distributions) may be problematic, and the mixture density approach presents opportunities to more accurately model null distributions. From the point of view of statistical theory, these methods more resemble estimation or modeling problems, while those based on p-value thresholds retain a closer connection to the structure of single hypothesis testing. These methods are discussed in Subheadings 4 and 5.

There exist a number of surveys of this topic, for example, (3, 4), with perhaps the most comprehensive treatment of the subject given in ref. (5). The aim of the current chapter is to provide an overview and classification of MTPs based on their underlying statistical issues, which will hopefully clarify their differences, thus aiding the reader in making a suitable selection.

2. Multiple Testing Procedures: Formal Definitions

At its simplest level, the objective of a MTP is to construct m indicator variables r_i which are close to τ_i. If the MTP sets $r_i = 1$, the hypothesis H_i is rejected, usually meaning that some phenomenon of interest has been detected associated with the ith component (usually a specific gene). If $r_i = 1$ but $\tau_i = 0$ this is referred to as a *false positive* or *false discovery*.

A common representation of the MTP is shown in Table 1. We assume $m_1 = \sum_i \tau_i$ hypotheses are false, and so $m_0 = m - m_1$ are true. The number of rejections is $R = \sum_i r_i$. Quantities m_0 and m_1 are unknown, hence the quantities U, V, T, and S are unobserved, but may be estimated. There will be special interest in the quantity $V^* = V / \max(R,1)$, interpretable as the proportion of rejected hypotheses which are truly null. The convention here is that no false discovery occurs when there are no rejections.

A MTP is characterized by several components. An *error rate* is an expected value of some function of the quantities of Table 1

Table 1
The table gives the structure of multiple hypothesis tests

Hypothesis type	Not rejected	Rejected	
True null hypotheses	U	V	m_0
False null hypotheses	T	S	m_1
	$m-R$	R	m

A total of m hypothesis tests are performed. In $m_0 = m - m_1$ of these, the null hypothesis represents the true state. The remaining quantities U, V, T, S are random entries in a two-way frequency table classified by test decision and true state

Table 2
Commonly used forms of error rate in MTPs, involving quantities defined in Table 1

Definition	Description
PCER $= E(V)/m$	Per-comparison error rate
PFER $= E(V)$	Per-family error rate
FWER $= P(V \geq 1)$	Family-wise error rate
FDR $= E(V^*)$	False discovery rate
pFDR $= E(V^* \mid R > 0)$	Positive false discovery rate

which serves as an measure of the accuracy of the inference. This is analogous to the level of significance of a single hypothesis test, except that more than one reasonable choice can be proposed. A MTP for which the error rate does not exceed λ *controls* for that rate at level λ. A distinction must be made, which does not arise in single hypothesis testing, between *strong control* (the error rate bound holds for any combination of true and false null hypotheses) and *weak control* (the error rate bound holds when $m_0 = m$, known as the *complete null hypothesis* CNH). Since we generally expect $m_1 > 0$, it is important to establish strong control of error rates.

A number of commonly used forms of error rates are listed in Table 2. Controlling by PCER is equivalent to treating each hypothesis test separately, with no multiple hypothesis adjustment. The FWER controls for the probability that at least one reported hypothesis rejection is incorrect, and has formed the basis for most MTPs before high-throughput applications. It interprets a report of rejected multiple hypotheses as a single inferential statement, which is or is not correct in its entirety. The appropriate form of error depends on the application, of course, but for high-dimensional screens, it would seem preferable to report the proportion of discoveries which are false (FDR), rather than attempt to control the probability of reporting any false discoveries. Such screens are exploratory in nature, and the cost in accepting a small number of false discoveries could be seen as smaller than the cost of forgoing true discovery, particularly when methods of post hoc discovery validation are often available.

The distinction between fixed and observed levels of significance extends to MTPs in a natural way. Given a fixed error rate λ for one of the quantities in Table 2, the MTP makes decisions r_i in such a way that the error rate holds. It should be noted that the error rate is an expected value applicable to a sample of MTPs. In the case of the FWER, the expected value is of an indicator variable, and thus does not differ from the single hypothesis test in this respect. However, the FDR is an expected value of a more complex random variable. It would be useful to know, for example, whether this random variable possesses especially heavy tails, yet these questions have not been widely considered in the literature.

If p_i is the original, or *unadjusted* p-value for H_i, the *adjusted* p-value p_i^a is defined as the minimum value of control λ at which the MTP sets $r_i = 1$. Level λ control is achieved by rejecting hypotheses for which $p_i^a \le \lambda$.

Unlike the single hypothesis test, the MTP also incorporates an estimation problem in the sense that $(r_1, \ldots, r_m)$ is an estimate of $(\tau_1, \ldots, \tau_m)$. Denote this rejection set

$$R = \{i : r_i = 1\}.$$

It is reasonable to insist that if $r_i = 1$, then $r_j = 1$ whenever $p_j \le p_i$. The MTP decision can then be reduced to the selection of a threshold t, calculable from $(p_1, \ldots, p_m)$ which yields rejection set

$$R(t) = \{i : p_i \le t\}. \tag{1}$$

2.1. Sequential Procedures: Single-Step, Step-Down, and Step-Up Forms

MTPs are generally designed to conform to Eq. 1, and most take one of three forms, the *single-step*, *step-up*, or *step-down* procedure, which are conveniently defined in terms of their adjusted p-values. Let $v_1, \ldots, v_m$ be a p-value anti-rankings, that is, $p_{v_1} \le p_{v_2} \le \ldots \le p_{v_m}$ (in the case of ties, the ranking need not be unique). The three forms are

$$\begin{aligned} p_i^a &= \min(C(m, p_i), 1) && \text{(single step procedure)} \\ p_{v_i}^a &= \max_{j \le i} \min(C(m, j, p_{v_j}), 1) && \text{(step-down procedure)}, \\ p_{v_i}^a &= \min_{j \ge i} \min(C(m, j, p_{v_j}), 1) && \text{(step-up procedure)} \end{aligned} \tag{2}$$

where the quantities $C(m, j, p)$ or $C(m, p)$ (referred to here as *adjustment functions*) are assumed to be (1) continuous increasing functions of p for fixed m, j and (2) nonincreasing functions of j for fixed m, p. We further assume that $C(m, j, 0) = 0$ and $C(m, j, 1) \ge 1$. These functions, along with the procedure form, define the particular MTP. Note that a single-step procedure can be regarded as either a step-down or step-up procedure in which the adjustment function $C(m, j, p) = C(m, p)$ is constant in j.

The best known single-step procedure is the Bonferroni procedure (B-SS), for which $C(m, p_i) = mp_i$.

$$p_i^a = \min(mp_i, 1), \quad i = 1, \ldots, m \quad \text{(B-SS)}.$$

Most remaining forms can be defined sequentially, with the following distinction. In a step-down procedure, we start with the smallest p-value, considering p-values in increasing order until we fail to reject a hypothesis. The rejection set is $R(t)$ where t is the value of the largest p-value rejected. In a step-up procedure, we start with the largest p-value, and consider p-values in decreasing order, stopping at the first p-value to be rejected. If that p-value equals t, the resulting rejection set is $R(t)$. The rejection rule for a p-value is given directly by $p_{v_i}^a \leq \lambda$, where λ is the reported error rate. For both the step-down and step-up procedure, the sequence $p_{v_i}^a$ is nondecreasing in i, so the structure is consistent.

2.2. Comparing MTPs: Conservatism and Ordering

A comparison of MTPs relies on a few concepts. An important consideration is whether or not the test is conservative. If so, the actual error rate is strictly smaller than λ, and for a given probability model this rate can, in principle, be calculated. In practice, this will generally not be possible, so some generality is desirable, even at the cost of some increase in conservatism. Another consideration is that sometimes a strict ordering between MTPs exists. If $p_{v_i}^a[A]$ and $p_{v_i}^a[B]$ are adjusted p-values from MTPs A and B we will claim the ordering $A \leq B$ whenever $p_{v_i}^a[A] \leq p_{v_i}^a[B]$ for all i. If a vector of p-values exists for which $p_{v_i}^a[A] < p_{v_i}^a[B]$ for some i, we have a strict ordering $A < B$. This type of relationship can exist between step-up and step-down procedures, by the following lemma.

Lemma 1

If A and B are step-up and step-down procedures using an adjustment function $C(m, j, p)$, then $A \leq B$. If there exists j^* for which $C(m, j^*, p) > C(m, j^* + 1, p)$ for any fixed m, $p \in (0, 1)$ then $A < B$.

Proof

Directly from the definitions, we have

$$p_{v_i}^a[A] \leq \min(C(m, i, p_{v_i}), 1) \leq p_{v_i}^a[B],$$

for each i, from which the ordering follows. Strict ordering is verified by considering any p-value vector approaching a vector of constant values p for which $C(m, 1, p) < 1$. Then $C(m, j^*, p_{v_{j^*}}) > C(m, j^* + 1, pv_{j^*})$, from which is follows that $p_{v_i}^a[A] < p_{v_i}^a[B]$ for any i.

Another form of ordering follows from a comparison of the adjustment functions.

Lemma 2

If A and B are step-down procedures using respective adjustment functions $C_A(m, j, p)$ and $C_B(m, j, p)$,and if $C_A(m, j, p) \le C_B(m, j, p)$ for all m, j then $A \le B$. Furthermore, if there exists j^* such that $C_A(m, j^*, p) < C_B(m, j^*, p)$ for all m, $p \in (0, 1)$ then $A < B$. This also holds if A and B are both step-up or both single-step procedures.

Proof

The ordering $\le$ follows directly from Eq. 2. We then investigate strict ordering for the three types of procedures separately.

Single-step procedure. Note that $C_A(m, j^*, p) < C_B(m, j^*, p)$ implies $C_A(m, p) < C_B(m, p)$, hence strict ordering follows directly from Eq. 2.

Step-down procedure. We may select a vector of p-values for which $p_{v_{j^*}} = p$, where $0 < C_A(m, j^*, p) < 1$. Then select all p-values of rank lower than j^* to be small enough so that $p^a_{v_{j^*}}[A] = C_A(m, j^*, p_{v_{j^*}})$. Then $p^a_{v_{j^*}}[B] = C_B(m, j^*, p_{v_{j^*}}) > C_A(m, j^*, p_{v_{j^*}})$, which completes the proof.

Step-up procedure. We may select a vector of p-values for which $p_{v_{j^*}} = p$, where $0 < C_A(m, j^*, p) < 1$. Then select all p-values of rank higher than j^* to be equal to 1. Then $p^a_{v_{j^*}}[A] = C_A(m, j^*, p_{v_{j^*}})$ and $p^a_{v_{j^*}}[B] = C_B(m, j^*, p_{v_{j^*}})$, which completes the proof.

This leads to the important special case.

Lemma 3

If A is a step-down procedure and B is a single-step procedure, with respective adjustment functions $C_A(m, j, p)$ and $C_B(m, p)$, then if $C_A(m, j, p) \le C_B(m, p)$ for all m, j, $p \in (0, 1)$ we have $A \le B$. If in addition $C_A(m, j^*, p) < C_B(m, p)$ for some j^* then $A < B$.

Proof

The lemma follows by interpreting B as a step-down procedure, then applying Lemma 2.

This leads to an important criterion for the selection of a MTP. If two MTPs A, B control for the same error rate, and $A \le B$, then there will be no advantage to choosing B. If $A < B$, then there is an advantage to choosing A.

2.3. p-Values with Ties

Finally, we say a brief word about p-value ties, which will occur when p-values are estimated by Monte Carlo methods such permutation tests, or bootstrap methods. It is important to verify that the MTP is coherent in the sense that any p-value is rejected if and only if all other p-values of the same value are rejected, and that any correct p-value ranking can be used. This is clearly the case with single-step functions. It turns out that no adjustment in the forms of Eq. 2 is necessary to ensure this for the other forms. The result is expressed in the following lemma, which follows directly from Eq. 2.

Lemma 4

Given a list of p-values $(p_1,\ldots,p_m)$, let j' and j'' be the minimum and maximum ranking of all p-values of value p^*. Assume control at rate λ. For a step-down procedure, all such p-values are rejected if and only if $C(m,j',p^*) \le \lambda$. For a step-up procedure all such p-values are rejected if and only if $C(m,j'',p^*) \le \lambda$.

Proof

The lemma follows from the fact that $C(m, j, p)$ is nonincreasing in j.

Since we have $C(m,j',p^*) \ge C(m,j'',p^*)$, we can see that step-up methods tend to be more rejective in the face of ties, an effect which will also hold to some degree when groups of p-values are otherwise very close in value.

2.4. Forms of Dependence

We assume that for each hypothesis test $i=1, \ldots, m$ there is an observed random vector $\tilde{W}_i$, often consisting of two samples of independent replications. To fix ideas, suppose we have two samples $\tilde{X} = (X_1,\ldots X_{n_1})$, $\tilde{Y} = (Y_1,\ldots Y_{n_2})$, each of which is a sample of m-dimensional random vectors. The data associated with dimension i is $W_i = (\tilde{X}_i, \tilde{Y}_i)$, $\tilde{X}_i = (X_{i1},\ldots,X_{in_1})$, $\tilde{Y}_i = (Y_{i1},\ldots Y_{in_2})$. In a typical microarray experiment, we have two independent samples of sizes n_1, n_2 of m-dimensional gene expression vectors distinguished by a phenotype. Each sample consists of replications which may be nested at various levels, ranging from the subject level to the technical level. Then X_{ij} (Y_{ik}) is the expression level of gene i for replication $j(k)$ from the first (second) sample. The concern for any MTP lies with the dependence structure across the gene index, that is, between W_i and W_j for genes i, j. Of some interest here is a discussion in (6), in which it is pointed out that gene expression assays are usually aggregated over many cells. Linear correlation is preserved over aggregation, but most other forms of dependence are degraded. Therefore, after suitable transformation linear correlation should be the dominant form of dependence.

It is widely recognized (see Chapter 10), that microarray gene expression profiles cannot be assumed to be independent. Despite this, MTPs requiring this assumption are commonly considered, not least for the very good reason that greater analytical precision is usually possible, and it is usually possible to shown that they remain valid under certain well defined dependence structures, either empirically or analytically. In other cases straightforward modifications to accommodate dependence exist.

It will be convenient to label dependence assumptions as they are encountered, starting with:

A(G): Any general form of dependence is permitted.
A(I): The p-values are statistically independent.

3. FWER: Refinements of the Bonferroni Procedure

Since the B-SS is probably the most commonly used MTP, a good way to clarify the issues behind the MTP is to review a number of refinements which can be used to control for FWER. The error rate is derived from Boole's inequality, which states that for any set of events $E_1,\ldots,E_K$:

$$P(U_{i=1}^{K}E_i) \leq \sum_{i=1}^{K} P(E_i). \tag{3}$$

The Bonferroni procedure follows by noting that if the probably of falsely rejecting a single hypothesis test is α, then applying Eq. 3 gives an estimate of FWER:

$$P(V \geq 1) \leq m_0\alpha \leq m\alpha,$$

which is obtained when tests are singly rejected at level α, which is therefore chosen to make $m\alpha \ll 1$. Since Eq. 3 holds for any sets, regardless of dependence structure, the B-SS itself must hold for all dependence structures, guaranteeing FWER no larger than $\lambda = m\alpha$ for the general case A(G). Inevitably, the cost of this generality is that the test is conservative in that the FWER is only an upper bound. It is important to note that this would be true even if m_0 were known, in which case we would report a FWER of $m_0\alpha$ (some methods of estimating m_0 will be discussed below).

3.1. Holm's Procedure: A Sharper Use of Boole's Inequality

In (7) the B-SS was modified, resulting in the *Holm's procedure* (H-SD), which is a step-down MTP with adjusted *p*-values given by

$$p^{a}_{v_i} = \max_{j\leq i} \min((m-j+1)p_{v_j},1), \quad i=1,\ldots,m \quad \text{(H-SD)}.$$

In (7), this procedure is referred to as the *sequentially rejective Bonferroni test*, but this description could be applied to a number of MTPs since developed). The H-SD controls for FWER and since, like the B-SS, it makes use only of Boole's inequality, it holds for assumption A(G). Furthermore, by Lemma 3 H-SD<B-SS, so there is no reason to use the B-SS when the H-SD is available.

The H-SD has been further refined in ref. (8).

3.2. Sidak's Inequality and Multivariate Distributions

Apart from such structural refinements of the B-SS, further refinement is possible by introducing specific dependence models. First suppose assumption A(I) holds. If we apply Taylor's theorem we get

$$\begin{aligned} P(V \geq 1) &= 1-(1-\alpha)m_0 \\ &= m_0\alpha - \frac{m_0(m_0-1)\alpha^2}{2} + \frac{m_0(m_0-1)(m_0-2)\alpha_r^3}{6}. \end{aligned}$$

for some $\alpha_r \in [0,\alpha]$. Since we assume $m\alpha \ll 1$ and if $m_0 \approx m$ the B-SS overestimates FWER in the A(I) case by a relative error of $O(m_0\alpha)$.

While it would be useful to recalculate FWER under assumption A(I), it has proven to be the case that a greater degree of generality can be obtained by instead considering alternatives to Boole's inequality. An important example is *Šidák's Inequality*, which we now state:

A(PODP): A set of random variables $U_1,\ldots,U_{m'}$ satisfy *Šidák's Inequality*, also referred at the *positive orthant dependence property* (PODP), if for any constants $c_1,\ldots,c_{m'}$

$$P\left(\prod_{i=1}^{m'}\{U_i \le c_i\}\right) \ge \prod_{i=1}^{m'} P(U_i \le c_i).$$

Then define the single-step *Šidák's Procedure* (Sk-SS):

$$p_i^a = \min(1-(1-p_i)^m, 1), \quad i=1,\ldots,m \quad \text{(Sk-SS)}.$$

Now suppose A(PODP) holds for a random vector $U_i : i \in T$, where U_i is a statistic associated with test H_i such that $\{U_i > c_i\}$ is a rejection region of size α. Then A(PODP) also holds for the *p*-values, and we may use the Sidák's inequality to obtain

$$\text{FWER} \le 1-(1-\alpha)^{m0} \le 1-(1-\alpha)^m.$$

This, of course, holds when the *p*-values are independent, but also under any dependence structure for which the inequality holds. It was shown in (9) that A(PODP) holds when $U_i = |Z_i|$, where $(Z_i,\ldots,Z_{m'})$ is a multivariate normal vector with marginal means equal to zero and any covariance matrix. Of course, the case of interest in the analysis of gene expression data is more likely to be the m'dimensional multivariate *t*-distribution $(T_1,\ldots,T_{m'})$ in which T_i is a *t*-statistic formed by the *i*th component from independent replications of $(Z_1,\ldots,Z_{m'})$, in which case $U_i = |T_i|$. Unfortunately, A(PODP) does not hold in general here. A counterexample is given in (10), where it was also shown that A(PODP) will hold when $\text{corr}(Z_i,Z_j) = \rho_i\rho_j$ for constants $\rho_1,\ldots,\rho_{m'}$. This holds, for example, when a random effect is added to each independent replicate of $(Z_1,\ldots,Z_{m'})$ with dimensional weighting permitted. A further generalization of this result is given in (11).

We then note that $1-(1-\alpha)^m \le \min(m\alpha,1)$ for $\alpha \in (0,1)$, with strict inequality for α>0. By Lemma 2 we have Sk-SS<B-SS, so that under assumption A(PODP) Sk-SS is preferable to B-SS. Furthermore, the single-step Sk-SS can be extended to *Šidák's Step-Down Procedure*, as shown in (12):

$$p_{v_i}^a = \max_{j \le i} \min(1-(1-p_{v_j})^{m-j+1}, 1), \quad i=1,\ldots,m \quad \text{(Sk-SD)}.$$

The Sk-SD controls for FWER under A(PODP), and by Lemma 3 we have Sk-SD<Sk-SS. Similarly, by Lemma 2 we have Sk-SD<H-SD. To summarize, under A(G) the best choice is H-SD, while under A(PODP) the best choice is Sk-SD.

We note that more detailed discussions of A(PODP) can be found in (12, 13).

3.3. Simes's Inequality

Another inequality of interest is *Simes' Inequality* (14):

A(SIMES): For ordered p-values $p_1,\ldots,p_{m'}$ and constant α the following inequality holds:

$$P\left(\bigcup_{i=1}^{m'}\{p_i \le \frac{i\alpha}{m'}\}\right) \le \alpha.$$

As originally proposed in (14) the procedure does not result in a rejection set R. We merely reject the CNH at level α if we observe $\bigcup_{i=1}^{m}\{p_{v_i} \le i\alpha / m\}$, assuming A(SIMES) holds under the CNH. A modification which permits statements regarding individual hypotheses is given in (15), and a step-up modification, *Hochberg's Step-Up Pro cedure* (Hg-SU) was proposed in (16):

$$p_{v_i}^{a} = \min_{j\ge i} \min((m-j+1)p_{v_j},1), \quad i=1,\ldots,m \quad \text{(Hg-SU)}.$$

Note that the Hg-SU differs from the H-SD only in that it is a step-up procedure, so by Lemma 1 we have Hg-SU < H-SD. But while H-SD holds under A(G), Hg-SU requires A(SIMES) to hold on the set of null hypotheses T. Under independence, Sime's inequality is exact for uniformly distributed p-values. In addition, empirical evidence in (14) shows that under a variety of practical models the estimate of FWER provided by A(SIMES) can be much more accurate than those based on Boole's inequality. The assumption A(SIMES) has been formally extended to a number of types of dependence structures. In (17) A(SIMES) was proven to hold for random vectors with common marginal densities satisfying the MTP2 condition:

A(MTP2): A random vector $X \in \Re^m$ with density f satisfies the MTP2 assumption if

$$f(x \vee y)f(x \wedge y) \ge f(x)f(y), \text{ for all } x, y \in \Re^m,$$

where $x \vee y = (\max(x_1,y_1),\ldots,\max(x_m,y_m))$ and $x \wedge y = (\min(x_1,y_1), \ldots, \min(x_m,y_m))$.

See refs. (17, 18) for a discussion of some common situations for which A(MTP2) holds, with the property extending to the p-values. As a point of comparison, we note that A(PODP) holds for any multivariate normal distribution, while A(MTP2) is shown in (16) only for positive exchangeable correlation.

Further refinements are proposed in (19, 20).

3.4. The Westfall and Young Procedure

Procedures which model dependence may take the general form proposed in (21), in which adjusted p-values are calculated in a single-step procedure using either of the two forms:

$$p_i^a = P_0(\min_{1 \le j \le m} U_j \le p_i) \text{ or } p_i^a = P_0(\max_{1 \le j \le m} |T_j| \ge t_i), \quad (4)$$

where $U_1,\ldots,U_m$ are distributed as the unadjusted p-values under the CNH, and $T_1,\ldots,T_m$ are distributed as the underlying test statistics under the same hypothesis, and t_i is the observed test statistic which results in p-value p_i. These procedures are usually referred to as the min P or max T single-step MTP. They provide strong control of FWER under the hypothesis of *subset pivotality* (21), which states that for any partition of hypothesis into truly and falsely null, the joint distribution of test-statistics associated with the true null hypotheses is the same as for the CNH. Note that Eq. 4 actually represents a class of MTPs. If Boole's inequality is used to estimate Eq. 4 the min P single-step procedures becomes equivalent to the B-SS, or to Sk-SS under A(I). The advantage of the formulation in Eq. 4 is that resampling techniques may be used to estimate the adjusted p-values, without requiring further assumptions about the dependence structure. The single-step procedure extends naturally to a step-down procedure using the quantities:

$$p_{v_i}^a = \max_{k=1,\ldots,i} P_0(\min_{j=k,\ldots,m} U_{v_j} \le p_{v_k}) \text{or } p_{v_i}^a = \max_{k=1,\ldots,i} P_0(\max_{j=k,\ldots,m} |T_{v_j}| \ge t_{v_k}), \quad (5)$$

which generally improves power over the single-step method.

3.5. Numerical Example

We will use a Monte Carlo study to illustrate some of the ideas of this section. We will consider three cases. In Case 1, $\tilde{X}$ and $\tilde{Y}$ are *iid* samples of sizes $n_1 = n_2 = 50$. We set X_1 to be a multivariate normal random vector of $m = 2{,}000$ dimensions, with the identity covariance matrix. The means of the first 100 coordinates are set to $\mu = 1$, the remainder to 0. Then Y_1 is constructed as X_1 except that all marginal means are set to 0. In Case 2, a random effect is introduced by setting $X_1 = X'_1 + kU$ and $Y_1 = Y'_1 + kU$, where X'_1, Y'_1 are equivalent to the Case 1 definition, k is a fixed vector of coefficients randomly selected from (–0.5,0.5), and U is a one-dimensional unit normal variate, selected independently for each replicate. For Case 3 X_1, Y_1 are defined as in Case 1, except that blocks of five coordinates possess a mutual correlation of –0.25.

To implement MTPs B-SS, H-SD, Sk-SS, Sk-SD, Hg-SU a two sample t-test was performed for each data set W_i, resulting in p-values $p_1,\ldots,p_{2{,}000}$. Note that Case 1 and Case 2 satisfy A(PODP) (Case 3 could be shown to satisfy A(PODP) is we used a normal density test with known variance). We also implemented a permutation-based implementation of min P and max T using the *multest* statistical library developed for the R statistical computing environment (22).

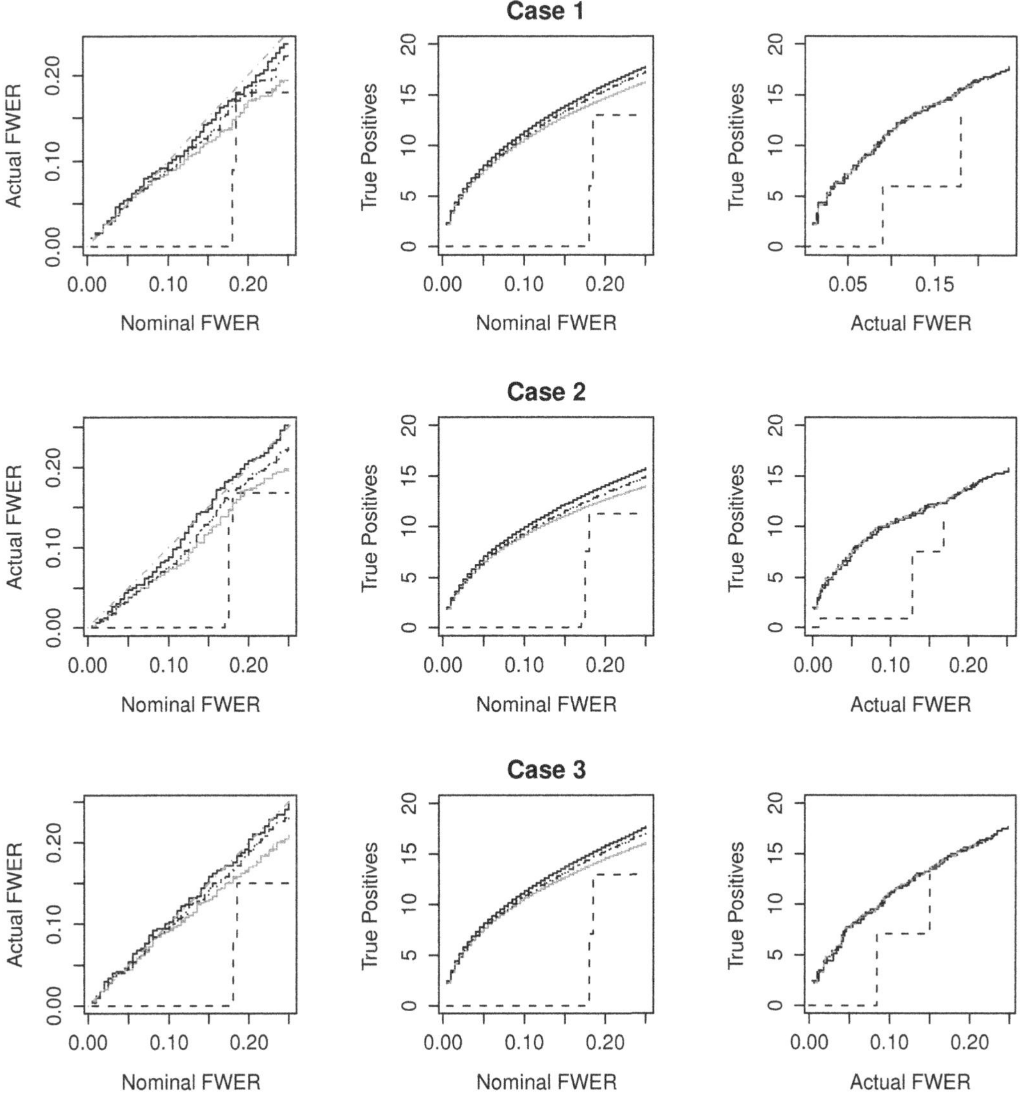

Fig. 1. Summaries of actual FWER, nominal FWER and power ($S=R-V$) for Cases 1, 2, 3 defined in Sub heading 2.4. Lines are (1) max *T* (*black solid line*), (2) min *P* (*black dashed line*), (3) B-SS (*gray solid line*), (4) H-SD (*gray dashed line*), (5) Hg-SU (*gray dotted line*), (6) Sk-SS (*black dotted line*), (7) Sk-SD (*black dash dotted line*), (8) Identity (*gray dash dotted line*).

A summary of the study is shown in Fig. 1. For each case the nominal FWER λ is compared to the estimated actual FWER (which can be observed given perfect knowledge of the underlying distribution). The average number of true positives $S=R-V$ (Table 1) is used as measure of the power of the procedures. As can be seen in the plots, the results are quite consistent across the three cases. First note that min P does not have sufficient resolution to compare with the other MTPs. Otherwise, a clear tendency emerges. In all cases, the max T procedure best estimates actual FWER, but without noticeably exceeding it. Of the remaining

MTPs, those based on Šidák's Inequality (Sk-SS, Sk-SD) better estimate FWER than do the the Bonferroni type MTPs. These tendencies then predict greater power, to the degree to which they are less conservative. In the third plot, the power is given as a function of the actual FWER, in which case there is little difference between the MTPs. This example therefore suggests that it is the correct estimation of FWER (or lack of conservatism) which determines the usefulness of the MTP.

4. The False Discovery Rate

While controlling for FWER prevents excessive spurious rejections, the very high dimension of high-throughput data has forced emphasis on the *false discovery rate* (FDR). This is appropriate when the cost of failing to report significant discovery can be balanced against the cost of reporting a small number of false positives, provided this error magnitude can be estimated.

4.1. Step-Up Procedures for FDR

A widely used application of the FDR was first proposed in (23) in the form of a step-up MTP, generally referred to as the *Benjamini-Hochberg Procedure* (BH-SU), which gives adjusted p-values:

$$p^a_{v_i} = \min_{j \geq i} \min\left(\frac{m}{j} p_{v_j}, 1\right), \quad i = 1, \ldots, m \quad \text{(BY-SU)}.$$

This procedure was originally shown to control FDR under assumption A(I), and was extended in (24) to a form of dependence known as *positive regression dependency* on a subset $I_0 \subset \{1, \ldots, m\}$ (PRDS).

A(PRDS): A set $D \subset \Re^m$ is an *increasing set* if $x \in D$ and $x \leq y \in \Re^m$ imply $y \in D$. A random vector $x \in \Re^m$, satisfies the PRDS property on $I_0 \subset \{1, \ldots, m\}$ if $P(X \in D \mid X_i = x)$ is nondecreasing in x for each $i \in I_0$ for any increasing set D.

If the test statistics for which p_i is an upper-tailed p-value satisfy A(PRDS) on the set of null hypotheses T, then BH-SU controls for FDR. It turns out that A(MTP2) implies A(PRDS).

Also in (24) a modification of the BH-SU is derived which controls FDR under $A(G)$, the *Benjamini-Yekutieli Procedure*:

$$p^a_{v_i} = \min_{j \geq i} \min\left(\frac{m\sum_{k=1}^{m} k^{-1}}{j} p_{v_j}, 1\right), \quad i = 1, \ldots, m \quad \text{(BY-SU)}.$$

For large m this is approximately equivalent to replacing the constant m in the BH-SU with $m\log(m)$.

4.2. Positive False Discovery

The pFDR (Table 2) is clearly similar to the FDR, but the differences have important implications for the form of inference used.

Under the CNH, we have $V=R$, hence $(V/R)I\{R>0\}=I\{V>0\}$, and FDR = FWER, while pFDR = 1, so that strong control of pFDR by a MTP is not strictly possible. Of course, pFDR = 1 has an intuitive explanation, namely that any rejections must be false positives. Treated as an estimation problem, if an estimate of pFDR is not significantly different from one, then we cannot reject the CNH. In addition, pFDR has a very natural quantitative interpretation. Returning to the mixture model introduced in Subheading 1, suppose we have two conditional densities $f_0(t)$ and $f_1(t)$. Suppose $\tau_1,\ldots,\tau_m$ can be interpreted as independent Bernoulli random variables with mean $\pi_1 = 1-\pi_0$. Then let $T_1,\ldots,T_m$ be an *iid* sequence defined by conditional density

$$T_i \mid \tau_i \tilde{}\, (1-\tau_i)f_0 + \tau_i f_1. \tag{6}$$

Integrating the conditional density Eq. 6 yields mixture density

$$f(t) = \pi_0 f_0(t) + \pi_1 f_1(t). \tag{7}$$

This generates the quantities in Table 1 in that hypothesis H_i is true if $\tau_i = 0$, T_i is the test statistic and f_0 is the null density. Suppose Γ is a common rejection region, and we write pFDR(Γ) to emphasize dependence on Γ. In (25) it was shown that under the assumption of independence, we have

$$\text{pFDR}(\Gamma) = P(\tau_i = 0 \mid T_i \in \Gamma) = \frac{\pi_0 P(T_i \in \Gamma \mid \tau_i = 0)}{P(T_i \in \Gamma)}, \tag{8}$$

so that pFDR(Γ) is interpretable as the Bayesian posterior probability that hypothesis H_i is true given that it was rejected, and π_0 is the prior probability. In (26) an analogue of the adjusted *p*-value for the pFDR is introduced. For a given statistic $T_i = t$ the *q-value* is defined as $q(t) = \inf\{\text{pFDR}(\Gamma)\text{: } \Gamma \text{ for which } t \in \Gamma\}$. Suppose the statistics T_i are expressed as *p*-values. Then f_0 is the uniform distribution, and rejection regions are of the form $\Gamma_\gamma = \{p : p \le \gamma\}$. The *q*-value for a given unadjusted *p*-value equal to *t* is defined as $q(t) = \inf_{\gamma \ge t} \text{pFDR}(\Gamma_\gamma)$. There is a technical issue worth noting here.

If pFDR(Γ_γ) is nondecreasing in γ, we have the simpler definition $q(t) = \text{pFDR}(\Gamma_t)$, but this condition does not necessarily hold. In (25) it is shown that it will hold when *p*-values are calculated for likelihood ratio tests.

4.3. Estimation of FDR and pFDR

The estimation of FDR or pFDR may follow from Eq. 8 in a natural way, and we briefly describe the approach proposed in (26). First suppose that the statistics T_i are *p*-values, so that we have $P(T_i \in \Gamma_\gamma \mid \tau_i = 0) = \gamma$. We then estimate $\hat{P}(T_i \in \Gamma_\gamma) = R_\gamma / m$ directly, where R_γ is the number of rejections using Γ_γ. As for π_0, a number of approaches are possible. We may simply note that $\pi_0 \le 1$, and for many high-throughput applications we will have $\pi_0 \approx 1$, so estimating π_0 as 1 yields a conservative estimate, at least when the variance of the other quantities is not taken into account. This expediency has been used in various contexts (23, 27). Alternatively, π_0 may be estimated using the concept of a *null dominant point*, that is, a point on the mixture density Eq. 7 at which the null component f_0 dominates. This concept is used in, for example, (28) to estimate properties of f_0. In (26) it is pointed out that for *p*-value densities, $t = 1$ will be a null dominant point, in which case we have $\pi_0 \approx f(1)$, which can be estimated from the data. This leads to an estimate $\hat{\pi}_0^{\lambda} = W_\lambda[(1-\lambda)m]^{-1}$ for some $\lambda \in [0,1)$, where $W_\lambda = \#\{T_i > \lambda\}$. Setting $\lambda = 0$ yields the conservative estimate $\hat{\pi}_0^0 = 1$. We finally note that the distinction between FDR and pFDR involves conditioning on $R_\gamma > 0$, leading to estimates

$$\begin{aligned} \text{p}\widehat{\text{FDR}}_\lambda(\gamma) &= \frac{{}^{W}\lambda\gamma}{\max(R_\gamma,1)(1-\lambda)[1-(1-\gamma)^m]} \\ \widehat{\text{FDR}}_\lambda(\gamma) &= \frac{{}^{W}\lambda\gamma}{\max(R_\gamma,1)(1-\lambda)}. \end{aligned} \tag{9}$$

A bootstrap procedure for selecting λ is proposed in (26). For further discussion of the equivalence of *q*-values and posterior error probabilities, see ref. (29).

5. Multiple Testing Procedures Based on Mixture Densities

While the FDR estimates of Eq. 9 rely on the mixture Eq. 7 for their definition, no actual modeling of this mixture is needed. On the other hand, if such a model did exist, we would have enough information to implement a MTP, as well as to gain further insight into the data not possible by the consideration of an error rate alone. In particular, such modeling permits flexibility in modeling of the null distribution.

We often know the form of f_0, at least up to a parameter, and can readily construct an estimate of f from the data, but this leaves the problem of estimating π_0 and f_1 indeterminant. However, some additional arguments may be introduced to mitigate this problem. The term *empirical Bayes* (30) used in this context refers to the estimation of quantities normally associated with prior distributions, in this case usually the quantities f_0 and π_0.

First, as noted in (30), interest may be in a posterior probability such as

$$P(\tau = 0 \mid T = t) = \frac{\pi_0 f_0(t)}{f(t)} \le \frac{f_0(t)}{f(t)}, \tag{10}$$

where τ is the hypothesis indicator. This has an estimable upper bound by noting $\pi_0 \leq 1$. A selection rule using this quantity as a threshold was shown in (30) to be equivalent to a conservative form of the BH-SU procedure.

It is then argued in (30) that the distinction between the theoretical density of f_0 and the empirical, or realized, density may be significant. Suppose we introduce the assumption that there is a null dominant point t_0. In (28) a scenario is studied in which f_0 is assumed to be approximately normal, but not exactly standard. The value t_0 is taken to be the mode of f_0, and the mean and variance of f_0 is then estimated by the location and curvature of that mode. This suffices to bound $P(\tau = 0 | T = t)$ using Eq. 10.

Alternatively, we may use p-values as the test statistic, in which case the theoretical value of f_0 is the uniform density. A null dominant point for this mixture would exist at $t_0 = 1$. This approach is taken in (31), in which f is modeled as a beta-mixture, with one component set as the uniform distribution with mixture parameter π_0. The problem of indeterminacy is resolved by assuming $f_1(1) = 0$.

5.1. Empirical Bayes in the Context of Multiple Testing

The relationship between the mixture formulation and control based on the FDR or the pFDR is quite close. In (30) the quantity

$$fdr(t) = \frac{\pi_0 f_0(t)}{f(t)} \tag{11}$$

is referred to as the *local false discovery rate* and is interpreted as the posterior probability that a hypothesis reporting statistic $T = t$ is a true null. We may then reject all tests for which $fdr(T_i) \leq \gamma$. This, by itself, is not a true MTP, but may easily be incorporated into the theory by noting the conditional expected value of $fdr(T)$ (taking T to be a p-value):

$$E[fdr(T) \mid T \leq t] = \pi_0 \int_{u \leq t} \frac{f_0(u)}{f(u)} \frac{f(u)}{P(T \leq t)} du = \frac{\pi_0 P(T \leq t \mid \tau = 0)}{P(T \leq t)} = \text{pFDR}(\Gamma_t)$$

by Eq. 8. Hence, the average local false discovery rate estimates $\text{pFDR}(\gamma_t)$:

$$\widehat{\text{pFDR}}(\Gamma_t) = \frac{\sum_{T_i \leq t} fdr(T_i)}{\#\{T_i \leq t\}}$$

and can be used to estimate the q-value $q(t)$. See ref. (32) for further discussion.

5.2. Fully Bayesian Models

It should be noted that although the methods described in Subheasdings 4 and 5 rely on Bayesian concepts, they are, strictly speaking, point estimates, whether parametric or nonparametric, and hence do not account for sampling variation. Fully Bayesian methods exist which are able to report modeling uncertainty and are more flexible when modeling dependence structure. They tend to be more specialized than the methods discussed here, and usually require more effort to implement. Despite this, where viable they form the most complete solution to the multiple testing problem.

Examples include (33, 34), in which hierarchical models are used to explicitly model the complete data across multiple differential expression experiments. Other interesting discussions can be found in (35–37).

6. Conclusion

The process of screening for differentially expressed genes using microarray samples can usually be reduced to a large set of statistical hypothesis tests. In this situation, statistical issues arise which are not encountered in a single hypothesis test, related to the need to identify the specific hypotheses to be rejected, and to report an associated error. As in any complex testing problem, it is rarely the case that a single method is always to be preferred, leaving the analysts with the problem of selecting the most appropriate method for the particular task at hand.

In this chapter, an introduction to current multiple testing methodology was presented, with the objective of clarifying the methodological issues involved, and hopefully providing the reader with some basis with which to compare and select methods.

References

1. Scherzer CR, Eklund AC, Morse LJ, Liao Z, Locascio JJ, Fefer D, Schwarzschild MA, Schlossmacher MG, Hauser MA, Vance JM, Sudarsky LR, Standaert DG, Growdon JH, Jensen RV, Gullans SR (2007) Molecular markers of early Parkinson's disease based on gene expression in blood. PNAS 104:955–960
2. Benjamini Y, Braun H (2002) John W. Tukey's contributions to multiple comparisons. Ann Stat 30:1576–1594
3. Yang YH, Speed T (2003) Statistical analysis of gene expression microarray data. In: Speed T (ed) Design and analysis of comparitive microarray experiments. Chapman and Hall, Boca Raton, FL, pp 35–92
4. Dudoit S, Shaffer JP, Boldrick JC (2003) Multiple hypothesis testing in microarray experiments. Stat Sci 18:71–103
5. Dudoit S, van der Laan MJ (2008) Multiple testing procedures with applications to genomics. Springer, New York, NY
6. Chu T, Glymour C, Scheines R, Spirtes P (2003) A statistical problem for inference to regulatory structure from associations of gene expression measurementswith microarrays. Bioinformatics 19:1147–1152
7. Holm S (1979) A simple sequentially rejective multiple test procedure. Scand J Stat 6:65–70
8. Shaffer JP (1986) Modified sequentially rejective test procedures. JASA 81:826–830
9. Šidák Z (1967) Rectangular confidence regions for the means of multivariate normal distribution. JASA 62:626–633
10. Šidák Z (1971) On probabilities of rectangles in multivariate Student distributions: their dependence on correlations. Ann Math Stat 42:169–175
11. Jogdeo K (1977) Association and probability inequalities. Ann Stat 5:495–504
12. Holland BS, Copenhaver MD (1987) An improved sequentially rejective rejective Bonferroni test procedure. Biometrics 43:417–423
13. Dykstra RL, Hewett JE, Thompson WA (1973) Events which are almost independent. Ann Stat 1:674–681

14. Simes RJ (1986) An improved Bonferroni procedure for multiple tests of significance. Biometrika 73:751–754
15. Hommel G (1988) A stagewise rejective multiple test procedure based on a modified Bonferroni test. Biometrika 75:383–386
16. Hochberg Y (1988) A sharper Bonferroni procedure for multiple tests of significance. Biometrika 75:800–802
17. Sarkar SK (1998) Some probability inequalities for ordered MTP_2 random variable: a proof of the Simes conjecture. Ann Stat 26:494–504
18. Sarkar SK, Chang C-K (1997) The Simes method for multiple hypothesis testing with positively dependent test statistics. JASA 92:1601–1608
19. Rom DR (1990) A sequentially rejective test procedure based on a modified Bonferroni inequality. Biometrika 77:663–665
20. Huang Y, Hsu JC (2007) Hochberg's step-up method: cutting corners off Holm's step-down methods. Biometrika 94:965–975
21. Westfall PH, Young S (1993) Resampling-based multiple testing. Wiley, New York, NY
22. Pollard KS, Dudoit S, van der Laan MJ (2005) Bioinformatics and Compu-tational Biology Solutions Using R and Bioconductor. In: Gentleman R, Huber W, Carey VJ, Irizarry RA, Dudoit S (eds) chapter Multiple testing-procedures: the multest package and applications to genomics (pp 249–271). Springer, New York, NY,
23. Benjamini Y, Hochberg D (1995) Controlling the false discovery rate: a practical and powerful approach to multiple testing. J R Stat Soc Ser B 57:289–300
24. Benjamini Y, Yekutieli D (2001) The control of the false discovery rate in multiple testing under dependency. Ann Stat 29:1165–1188
25. Storey JD (2003) The positive false discovery rate: a Bayesian interpretation and the *q*-value. Ann Stat 31:2013–2035
26. Storey JD (2002) A direct approach to false discovery rates. JSS-B 64:479–498
27. Efron B (2003) Robbins, empirical Bayes and microarrays. Ann Stat 31:366–378
28. Efron B (2004) Large-scale simultaneous hypothesis testing: the choice of a null hypothesis. JASA 99:96–104
29. Kall L, Storey JD, MacCross MJ, Noble WS (2008) Posterior error probabilities and false discovery rates: two sides of the same coin. J Proteome Res 7:40–44
30. Efron B, Tibshirani R, Storey JD, Tusher V (2001) Empirical Bayes analysis of a microarray experiment. JASA 96:1151–1160
31. Allison DB, Gadbury GL, Moonseong H, Fernandez JR, Cheol-Koo L, Prolla TA, Weindruch R (2002) A mixture model approach for the analysis of microarray gene expression data. Comput Stat Data Anal 39:1–20
32. Newton MA, Wang P, Kendziorski C (2006) Hierarchical mixture models for expression profiles. In: Do K, Muller P, Vannucci M (eds) Bayesian inference for gene expression and proteomics. Cambridge University Press, New York, NY, pp 40–52
33. Newton MA, Kendziorski CM, Richmond CS, Blattner FR, Tsui KW (2001) On differential variability of expression ratios: improving statistical inference about gene expression changes from microarray data. J Comput Biol 8:37–52
34. Newton MA, Noueiry A, Sarkar D, Ahlquist P (2004) Detecting differential gene expression with a semiparametric hierarchical mixture method. Biostatistics 5:155–176
35. Lewin A, Richardson S, Marshall C, Glazier A, Aitman T (2006) Bayesian modeling of differntial gene expression. Biometrics 62:1–9
36. Gottardo R, Raftery AE, Yeung KY, Bumgarner RE (2006) Bayesian robust inference for differential gene expression in microarrays with multiple samples. Biometrics 62:10–18
37. Do K, Muller P, Vannucci M (2006) Bayesian inference for gene expression and proteomics. Cambridge University Press, New York, NY

Chapter 4

Gene Selection with the δ-Sequence Method

Xing Qiu and Lev Klebanov

Abstract

In this chapter, we discuss a method of selecting differentially expressed genes based on a newly discovered structure termed as the δ-sequence. Together with the nonparametric empirical Bayes methodology, it leads to dramatic gains in terms of the mean numbers of true and false discoveries, and in the stability of the results of testing. Furthermore, its outcomes are entirely free from the log-additive array-specific technical noise. The new paradigm offers considerable scope for future developments in this area of methodological research.

Key words: Microarray data, Correlation, Differential expression, Gene pairs

1. Introduction

One of the main focus of using microarray data analysis is to select genes that are *differentially expressed* between two (or more) biological conditions (phenotypes). The rationale behind this practice is that if a gene is biologically important in determining the different phenotypes, this importance should be reflected from the difference between the distribution functions of this gene in different biological conditions.

Most currently practiced methods are designed to select those genes of which the *marginal distributions* are different between two biological conditions. Among these methods, most focus solely on detecting the changes of the *means* of the marginal distributions.

The main obstacle to analyzing microarray data is the high dimensionality of the data. The number of microarrays in a study is typically very small (more than often it is <10 for each phenotype) when compared to the number of hypothesis to be tested (ranging from thousands to tens of thousand typically). Sometimes it is called the "large p, small n" paradigm.

Andrei Y. Yakovlev et al. (eds.), *Statistical Methods for Microarray Data Analysis: Methods and Protocols*, Methods in Molecular Biology, vol. 972, DOI 10.1007/978-1-60327-337-4_4,

The "large p" nature of this type of analysis calls for *multiple testing adjustment*. Due to the extreme large number of hypothesis, a loose significance level for single hypothesis testing can result in appallingly large number of false positives. The following example is simple yet enlightening.

Example 1

We have collected gene expression level data from 12 individual blood samples (12 slides). Six of them are drawn from healthy individuals, six from patients with prostate cancer. Each slide contains (log-transformed) expression level information of 20,000 genes. Furthermore, let us assume all but ten genes are differentially expressed.

If we choose to not apply any multiple testing procedure, there will be about 1,000 false positives on average when the significance level is 0.05. This is why multiple testing adjustments are so important.

Typically, the gene selection methods can be summarized in the following three steps:

1. Apply a suitable univariate test for each gene. Among the more popular choices are Student's t-test and its derivatives, or its nonparametric counterparts such as the Wilcoxon rank-sum test. The results of this step are (unadjusted) p-values associated with each gene.
2. Apply a multiple testing procedure (MTP) to those unadjusted p-values to obtain *adjusted* p-values. The choices of MTPs include the old faithful Bonferroni procedure and another half dozen of its cousins which aim at controlling the familywise error rate (FWER); the Benjamini-Hochberg procedure and its friends which aim at controlling the false discovery rate (FDR).
3. Select a threshold so that genes with adjusted p-values less than this cutoff value will be declared as differentially expressed genes.

These methods will be called collectively as the *univariate selection methods* henceforth. From a naïve point of view, these methods are very different and some of them seem to be much more powerful than the others. For example, given a particular dataset and a given nominal threshold, different univariate selection methods can produce list of differentially expressed genes with different lengths, ranging from just a few to thousands. Naturally, practitioners tend to call those which selects more genes to be more powerful. This "common sense" is wrong because different MTPs are designed to control different statistical quantities. It is unfair to compare a procedure which controls FWER at level 0.05 to a procedure which controls FDR, also at the same 0.05 level.

Instead, we may ask a more fundamental question: do these MTPs change the *ordering* defined by unadjusted p-values? The answer is "no" for a large class of MTPs. In other words, if the unadjusted p-value associated with gene g_1 is smaller than that of gene g_2, the adjusted p-value of g_1 calculated from one of the above MTPs will be smaller than that of g_2 as well. Consequently, if a given univariate selection method declares g_2 to be differentially expressed, g_1 must be in the list too. In this sense, all the "tricks" are equivalent to assigning different thresholds for unadjusted p-values. Furthermore, the ROC curves of these methods are either identical or near identical to that of the selection method based on the unadjusted p-values. Interested reader may refer to (26) for more details.

Let us further assume that all log-transformed expression levels in Example 1 are normally distributed with the same variance σ^2. The effect size (difference in mean) for each one of the ten differentially expressed genes is one σ. The univariate test to be used is two sample t-test.

It can be easily shown that in order to select 1 out of 10 differentially expressed genes on average (power = 0.1), the significance level for the two sample t-test should be 0.0066 for the unadjusted p-values. At this significance level, we would make approximately 132 false selections on average. One either have to accept more than 132 false positives on average, or <1 out of 10 true discoveries on average. The more depressing fact is that this dilemma will not be solved by any order preserving MTP.

This dire situation is caused by the "small n" nature of microarray data, and it cannot be solved without increasing the sample size dramatically under the current univariate selection method framework.

As a side note, all univariate selection methods are biased towards selecting genes that display the most pronounced univariate difference (difference in its marginal distribution). However, many phenotypic changes are triggered by a modest change in expression of a modulator gene, which plays a *catalytic* role in a specific cell function. The small change of this gene can be amplified by a long chain of downstream genes, so that these genes are more likely to be selected by the univariate selection methods. As we have discussed, most MTPs do not change this order. Thus thousands of irrelevant downstream genes have to be selected, or the really important modulator gene will be missed. This predicament can be solved only by well-designed selection procedures which exploit changes of the *joint distribution* of genes expression.

2. Nonparametric Empirical Bayes Methodology

To overcome the limitation of small sample size, many methods have been developed based on the idea of "borrowing power from other genes," or pooling information across genes. One notable example is the nonparametric empirical Bayes methodology (see refs. (1–3)) abbreviated as NEBM.

NEBM starts with a simple Bayesian model: there are two classes of genes, "differential expressed" (**DE**) and "not differentially expressed" (**NDE**). This difference is characterized by the different density functions of a suitable test statistic *T*. Denote $f_1(t)$ and $f_0(t)$ as the density functions of this test statistic for the **DE** class and the **NDE** class. Here the role of *T* is to summarize the information contained in the data. No *p*-value is computed based on this statistic and no parametric assumption is needed, thus any ***pivotal*** test statistic can be used. As an example, two sample Student's *t*-statistic is a good choice.

Let the prior probabilities for the **DE** class and the **NDE** class be $1-\pi_0$ and π_0, respectively. Then the density function for the mixture density, $f(t)$, is:

$$f(t) = \pi_0 f_0(t) + (1-\pi_0) f_1(t). \tag{1}$$

According to Bayes Theorem, ***a posteriori*** probability for a gene to be in the **DE** class given its summary statistic $T=t$ is:

$$p_1(t) = \Pr(\mathbf{DE} \mid T = t) = 1 - \pi_0 \frac{f_0(t)}{f(t)}. \tag{2}$$

We can estimate $p_1(t)$ by

$$\hat{p}_1(t) = 1 - \hat{\pi}_0 \frac{f_0(t)}{\hat{f}(t)}, \tag{3}$$

where $\hat{\pi}_0, \hat{f}_0$ and $\hat{f}(t)$ are some suitable estimators for π_0, f_0, and $f(t)$.

Differentially expressed genes can be selected by a simple Bayes rule: a gene is said to be differentially expressed if its associated $p_1(t)$ is greater or equal to a prespecified threshold level *C*, such as $C=0.8$.

There are two ways to deal with the prior probability π_0. The first one is to use the most conservative choice $\pi_0 = 1$, so that

$$\hat{p}_1(t) = 1 - \frac{f_0(t)}{\hat{f}(t)}. \tag{4}$$

This approach is appropriate if we have some empirical evidence to believe that only a small proportion of genes are differentially expressed.

Another approach is to estimate π_0 from the data. This topic is mentioned in many papers (4–11).

There are two reasons that we prefer the first, more conservative approach.

1. The prior probability π_0 is unidentifiable without parametric assumptions on the densities $f(t)$ and $f_0(t)$ (1). This defeats one of the major advantage of NEBM, that is, it does not depend on any specific parametric modelling assumption.

2. Estimating π_0 adds additional instability and variability to the NEBM (12, 13), which is the main disadvantage of this otherwise elegant method.

There are several different choices of $f(t)$. The simplest nonparametric estimate of this density function is its histogram. However, it turns out that the histogram density estimator is highly variable, thus the set of differentially expressed genes selected by the NEBM built upon it are highly unstable. In (1), Efron suggests that Poisson GLM fit should be used as a smoothing technique to reduce this variability. Kernel estimation (with Gaussian kernel, for example) is an alternative smoothing technique. Compare to the Poisson GLM fit, this method is more flexible and computationally much less expensive. According to the simulations and studies involving real biological data (12), its performance is comparable to that of the Poisson GLM fit.

Using Student *t*-statistics as the summary statistic T and assuming normality of the log-transformed expression levels, T would follow the Student *t*-distribution. However, there is no clear evidence to show that gene expressions are normally distributed. In this case, we can estimate $f_0(t)$ by means of (unbalanced) permutation method. Suppose there are n_1 and n_2 microarray slides in each phenotype, each containing information about m genes. First randomly permute all slides ($n_1 + n_2$ in total), then split them into two groups again of size n_1 and n_2. Summary statistics will be computed and recorded for this pseudo dataset. We repeat this step for K times, then either the Poisson GLM fit or the kernel smoothing is employed to provide $f_0(t)$. Efron (1) suggests that $K=20$.

If $n_1 = n_2$ and they happen to be even numbers, we may apply balanced permutation: randomly select $n_1 / 2$ slides from the first phenotype and move to the second phenotype and vice versa. However, this method results in a much higher variance of the true/false positives (12). So even for this case, we still recommend using unbalanced permutation.

The NEBM can be summarized in the following steps:

1. Choose a suitable summary statistic and compute this statistic for each gene.
2. Estimate $f(t)$ from these statistics.
3. Use unbalanced permutations to model the log-transformed expressions under the null hypothesis.
4. Estimate $f_0(t)$ from the statistics obtained from these permutations.
5. Compute p_1, the estimated *a posteriori* probability by Formula 4.4.
6. Declare genes with $p_1 \geq C$ as being differentially expressed.

The NEBM has some distinct attractive features. In particular, the NEBM allows one to work with any pivotal test statistic (including continuous statistics) without any distributional assumptions, as long as the test statistics are near independent and identically distributed under the null hypothesis. This in turn leads to a higher power in comparison to rank-based distribution-free tests.

However, this method suffers from high variability of its results, to the extent that the results can hardly be trusted. This is due to the fact that gene expression levels, and consequently the summary statistics are far from being independent, so the density estimates $f_0(t)$ and $f(t)$ are highly variable and even inconsistent. To obtain theoretical convergence results, some researchers assume the weak dependence between the gene expression levels (14, 15). This hypothesis is based on such believe: genes tend to work in small groups (pathways) which involves just a few to 50 or more genes, and each group works independently. However, an empirical study (16) suggests such groups involve thousands of tightly dependent genes.

Recently, Klebanov et al. (17) discovered a new type of stochastic dependence between expression levels in gene pairs. Termed as the Type A gene pairs, this modulation like unidirectional dependence between expression signals arises when the expression of a "gene-modulator" stochastically proportional to that of a "gene-driver." Not only there are abundant gene pairs (over 35% of all pairs) fall into this category, there are genes that tend to form the Type A relationships with the overwhelming majority of genes. This new discovery shows how diverse the correlation structure could be, and how erroneous the assumption of independence or weak independence is.

2.1. δ-Sequence, A Structure Yielding Near-Independent Random Variables

As discussed in the previous sections, NEBM is highly unstable due to the long-ranged correlations between log-transformed expression levels and, consequently, between the associated test statistics (16, 18). NEBM can be seen as an effort to overcome the "curse of dimensionality" in microarray data analysis by "borrowing strength from other genes." When making inference of changes in expression level of a particular gene, it needs information of all other genes to construct density curves. While the information contained in a single observation is always enriched by adding another independent and identically distributed observation, this is not quite so for dependent observations. As an extreme example, if all variables are deterministically dependent, the information contained in one of them cannot be further enriched by any additional observation. The situation is exacerbated when the observations have dissimilar distributions. In earlier publications, abundant evidence (12, 19, 20) shows that the variability of the results of those methods that resort to pooling across genes is intolerably high. The consequences of this instability can be disastrous, as one may frequently declare thousands of genes as differentially expressed while none of them are true discovery.

This dire situation drove us to search for a structure in the form of a long sequence of suitably *transformed variables* which are *near independent*. This would put us in the desirable situation where methods that resort to pooling test statistics across genes, such as the adaptive FDR-controlling procedures (8, 21), the non-parametric empirical Bayes method (1–3), or some robust statistical methods, may be very advantageous.

To accomplish this, we exploit a special structure found in gene expression signals termed as the δ-sequence. This structure was first discovered in a set of microarray data reporting log-transformed expression levels of $m=7{,}084$ genes in $n=88$ patients with hyperdiploid acute lymphoblastic leukemia (HYPERDIP data, Affymetrix GeneChip platform, St. Jude Children's Research Hospital Database (22)). The presence of the δ-sequence has been confirmed in several other data sets as well.

3. Construction of δ-Sequence

We start with ordering genes with their standard deviation in increasing order. Denote the log-transformed expression levels of those gene as X_i, $i=1, \ldots, m$. X_1 has the smallest standard deviation, and X_m has the largest standard deviation. Then we form a sequence of the increments $\delta_1 = X_2 - X_1, \ldots, \delta_{m/2} = X_m - X_{m-1}$. This sequence will be called the δ-sequence henceforth. Notice that m is an even number in our case. If the total number of genes is odd, the last (or the first) gene in the original sequence can be discarded.

The main property of the δ-sequence is that its components are near independent. Figure 1 compares the histogram of sample correlation coefficients between log-transformed expressions and between δ_i. A closer study of the histogram of Fisher's transformation *z*-scores computed for all pairs of δ_i reveals that δ_i are not entirely independent. Residual correlation, albeit small, exists between δ_i. This is the reason we call the δ-sequence *near independent*.

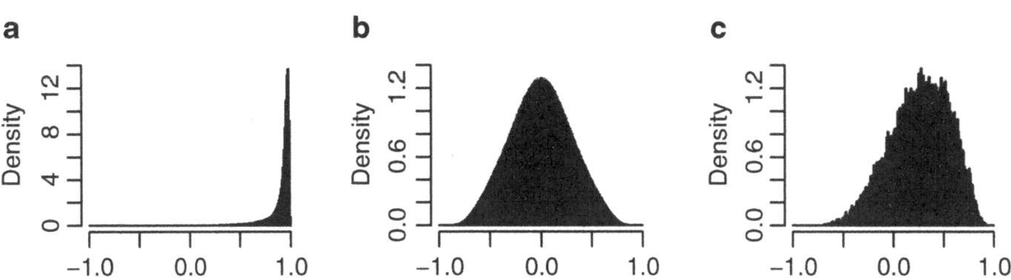

Fig. 1. Histogram of all pairwise correlation coefficients. (**a**) log-transformed expression levels; (**b**) δ-sequence; (**c**) large–small STD pairs. Data being used: HYPERDIP.

Denote the distribution function of δ_i of two different phenotypes as $F_{\delta_i(A)}(x)$ and $F_{\delta_i(B)}(x)$, respectively. Since δ_i are linear combinations of gene pairs, rejecting the following hypothesis

$$\mathbf{H}'_i : F_{\delta_i(A)}(x) = F_{\delta_i(B)}(x), \tag{5}$$

implies that at least one of the members of gene pair $(2i-1, 2i)$ must be differentially expressed. This enables us to apply a gene selection method, such as the NEBM on δ_i to select differentially expressed gene pairs, and then use some method to purge those genes that are not differentially expressed from these pairs.

One simple ad hoc method to accomplish this is called the shift method.

1. Select differentially expressed nonoverlapping gene pairs by applying the NEBM to the δ-sequence.
2. Keeping the same gene ordering defined by the sample standard deviation, shift the starting point to the second gene in the ordering and form another δ-sequence. In other words, we define $\delta_i = X_{2i+1} - X_{2i}$ instead. The first gene can be paired with the last gene if m is an even number, or it can be discarded if m is an odd number. Using the newly formed alternative δ-sequence to select differentially expressed genes in nonoverlapping pairs by the NEBM.
3. Report the intersection of the two gene sets thus selected as the final list of candidate genes.

It turns out that this simple method works very well. However, there are still some room left for improvement in this direction. We will come back to this topic in the next section.

4. The Performance of This New Method

The discovery of the δ-sequence makes it possible to remove the main obstacle standing in the way of applying the NEBM to microarray data. It can also be applied to other multiple testing problems where the dependence between tests is seen as a major nuisance.

To best understand the advantage of using the NEBM with the δ-sequence, we conduct the following simulation studies based on resampling from the HYPERDIP data.

Prior to subsampling, the genes were ordered by the sample standard deviation of their log-transformed expressions. 350 randomly selected genes were selected and will be assigned some artificial differences later. All 88 arrays were used to estimate the sample standard deviation. The initial ordering and the set of the selected 350 genes were kept the same throughout the resampling experiment.

Table 1
Comparing δ-sequence method to expression level-based methods

	Mean false positives				Mean true positives			
n	n/n	Global	Quantile	Deltas	n/n	Global	Quantile	Deltas
5	1184.72	55.33	62.20	19.09	8.38	305.18	310.86	213.64
10	826.55	68.67	88.71	18.68	10.65	338.33	345.39	279.85
20	588.82	81.36	133.09	18.19	35.35	347.17	349.76	306.80
30	482.76	97.50	188.30	17.61	92.65	349.14	350.00	313.10
40	400.44	123.61	252.24	17.50	176.33	349.75	350.00	315.88

n/n: no normalization

At every step of the resampling procedure, two subsamples of subjects (slides), each of size n, were randomly drawn without replacement from the HYPERDIP data ($n = 88$). The choices of n are 5, 10, 20, 30, and 40. One subsample was modified by adding effect sizes to the 350 preselected genes on each slide, while the other subsample was left intact. This is a natural way to model shift alternatives, so the 350 preselected genes forms the **DE** group while the **NDE** group consists of the other genes. To make such effects reasonably invariant across different scales of the data, for each differentially expressed gene, we set the effect size to be one standard deviation of its expression level (estimated from the original sample, $n = 88$).

It should be kept in mind that this way of simulating the true alternatives is not absolutely perfect, as it amounts to requiring the condition of subset pivotality (23) to be met in the data under consideration. However, this is probably as far as we can go with resampling in an effort to preserve the actual joint distribution of the biological data.

For gene selection, NEBM based on t-statistic was used with a conservative estimate $\pi_0 = 1$. The null and alternative density functions were estimated by kernel smoothing with Gaussian kernel. For the null distribution, unbalanced permutation method was used. The posterior probability cutoff was chosen to be 0.9. The procedure was repeated 3,000 times, the number of false/true positives were recorded.

A similar study was conducted by using the NEBM without the δ-sequence. The original log-transformed expressions as well as their normalized values were used. The two normalization procedures being used are: (1) the global normalization (24), (2) the quantile normalization (25).

Tables 1 and 2 provide some information about the performance of the δ-sequence method comparing to that of the expression level-based method.

Table 2
Comparing δ-sequence method to expression level-based methods

	STD of false positives				STD of true positives			
n	n/n	Global	Quantile	Deltas	n/n	Global	Quantile	Deltas
5	1843.22	103.12	49.08	18.20	49.39	25.63	23.95	31.43
10	1882.91	143.71	68.72	10.62	56.58	8.29	3.17	14.16
20	1616.77	142.76	92.71	11.00	102.63	3.30	0.50	8.90
30	1442.55	173.31	115.74	6.83	152.04	1.65	0.07	7.89
40	1338.64	199.59	140.60	7.01	172.12	0.80	0.00	7.59

n/n: no normalization

From Table 1, we see that the NEBM with the δ-sequence produced much fewer mean false positives compared to other methods. For example, when there were 40 slides in each group, the mean number of false positives produced by the δ-sequence method was 17.50, which was one magnitude of order less than any other alternatives. The δ-sequence method was even more advantageous in terms of the stability of false positives, which is made evident by Table 2. Another point is worth noting: like the non-normalized data, when the sample size became larger, the mean and the standard deviation of false positives decreased with the δ-sequence method. It was exactly the opposite for the two normalization methods.

It has to be noted that these nice properties of the δ-sequence method came at a small sacrifice of power, indicating that there were some information lost in the pairing of genes.

The success of the δ-sequence method lies in the fact that

- It is an efficient way to produce near-independent sequence of surrogate variables.
- These surrogates are gene pairs, as oppose to combinations of thousands of genes or even the whole genome required by other decorrelation methods such as PCA (principal components analysis).

Recall that the construction of the δ-sequence can be summarized in three steps:

1. Sort genes according to their standard deviation.
2. Divide the list of genes into consecutive nonoverlapping pairs.
3. Take the differences of these gene pairs.

Why this simple procedure works so well in reducing the correlation coefficients?

Figure 1 shows the histograms of the pairwise correlation coefficients in three different situations. Figure 1a is the histogram of the pairwise correlation coefficients between the log-transformed expressions. It should that most of the pairwise correlation coefficients locate in a small neighborhood of one. Figure 1b shows that the pairwise correlation coefficients between δs are much lower. This histogram is symmetric around zero. Figure 1c is a specially designed experiment, where we took 200 genes with the largest observed standard deviation to pair with genes with the smallest observed standard deviation, and then take the difference. Compare Fig. 1c to Fig. 1b, we see that pairing genes with the order defined by the observed standard deviation results in much smaller pairwise correlation coefficients.

In fact, sorting genes according to their observed STD is useful because it makes the two genes in the same pair, denoted as x_1 and x_2, have similar STD.

Let (Ω, F, P) be a probability space and denote $L^2(\Omega,\mathcal{F},P)$ to be the Hilbert space of random variables with finite second order moment. The inner product $L^2(\Omega,\mathcal{F},P)$ in is given by $< x_1, x_2 >= \int_{\Omega} x_1 x_2 dP$. Denote $\mathbf{R}^1$ to be the space of constant random variables. Because $\mathbf{R}^1$ is a one-dimensional subspace of $L^2(\Omega,\mathcal{F},P)$, we can define a quotient space $\mathcal{H} = L^2(\Omega,\mathcal{F},P) / \mathbf{R}^1$. The members of $\mathcal{H}$ are equivalence classes of random variables which are identical up to a constant. It is clear that $\mathcal{H}$ is isomorphic to the Hilbert space of random variables with zero mean, with $< X, Y >= \mathrm{cov}(X, Y)$. For convenience purpose, we use the same notation x to denote the equivalence classes of which x is a member.

In plain words, covariance can be considered as an inner product between two random variables so long as the constant term of a random variable can be ignored. Following the same reasoning, $STD\,(x_1)$ can be considered as the length of x_1. The arc-cosine of correlation coefficient between x_1 and x_2 can be thought as the angle between these two random variables. We know that the log-transformed expression levels of genes are highly correlated, so x_1 and x_2 can be visualized as *near parallel* vectors in $\mathcal{H}$. Now if x_1 and x_2 have comparable length, the difference $\delta = x_1 - x_2$ will be almost perpendicular to both x_1 and x_2. δ is almost perpendicular to other genes as well because other genes are near parallel to both x_1 and x_2. If $|x_1|$ is much greater than $|x_2|$, then the difference $x_1 - x_2$ points to a direction that is near parallel to that of x_1. Below is an example to illustrate this difference:

Example 2

Let z, w be two independent random variables with standard normal distribution. Let $x_1 = z + w$, $x_2 = 0.5z + 1.3w$, and $x_3 = 2.5z + 3.0w$.

It is easy to show that $STD(x_1) = 1.414$, $STD(x_2) = 1.393$, $\mathrm{Corr}(x_1, x_2) = 0.914$. So x_1, x_2 are highly correlated, and they have similar standard deviations. The standard deviation of x_3 is 3.905,

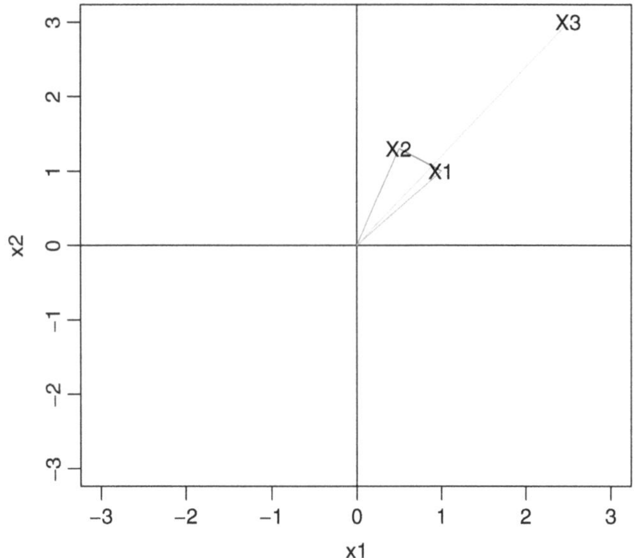

Fig. 2. δ_1 (the red line segment) is almost perpendicular to X_3 while the angles between and X_3 are very acute.

Corr(x_1, x_3) = 0.996, Corr(x_2, x_3) = 0.933. In other words, x_3 are also highly correlated with both x_1 and x_2. Note that $STD(x_3)$ is much greater than both $STD(x_1)$ and $STD(x_2)$.

Denote $\delta_1 = x_2 - x_1 = -0.5z + 0.3w$, we have $STD(\delta_1) = 0.583$, $\text{Corr}(\delta_1, x_3) = -0.154$. So the correlation between δ_1 and x_3 is significantly smaller than Corr(x_1, x_3) and Corr(x_2, x_3). This can be seen from Fig. 2.

Now define δ_2 to be the difference between x_3 and x_1, $\delta_2 = x_3 - x_1 = 1.5z + 2.0w$. It is easy to compute that the correlation between δ_2 and x_2 is $\text{Corr}(\delta_2, x_2) = -0.962$, even greater than both $\text{Corr}(x_1, x_2)$ and $\text{Corr}(x_3, x_2)$. This can be seen from Fig. 3.

It turns out that in terms of correlation reduction, the δ-sequence method works the best under this model:

$$x_i = c_i\alpha + y_i.$$

As before, x_i denotes the log-transformed expression level of the ith gene, α is a hidden factor shared by all genes. c_i is the weight of this factor for the ith genes. We assume these c_i are dense in the sense that if you order them from the smallest to the largest, the difference between any two consecutive c_i is small. The last term, y_i represents information specific to each gene.

Furthermore, we assume

1. $\text{Corr}(x_i, x_j) \approx 1$.
2. $\text{Corr}(\alpha, y_i) = 0$ for all i.
3. $\text{Corr}(y_i, y_j) = 0$ for all $i \neq j$.

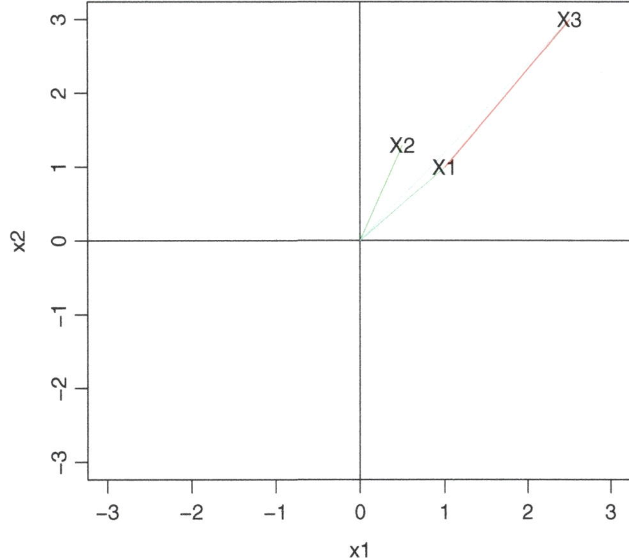

Fig. 3. δ_2 (the *red line* segment) is almost parallel to X_1, X_2 and X_3.

It can be shown that under these simplistic assumptions, δ_i are essentially uncorrelated to each other.

It should be emphasized that correlation reduction is not the only criterion to assess the utility of the δ-sequence method or other related variable transformation methods. After all, our goal is to accurately find *individual genes* that behaves differently between biological conditions. A variable transformation method that yields more independent surrogate variables, but fails to retain the information of individual genes cannot be a good method. As an example, the eigenvectors of the covariance matrix (as used in the principal component analysis) are uncorrelated to each other, but they are not good candidates to be used in conjunction with the NEBM or any other gene selection procedure for an obvious reason: these eigenvectors are linear combinations of *all variables* (genes). It is next to impossible to interpret "differentially expressed eigenvectors of genes," should we be luck to find any at all.

It is known that various normalization procedures can reduce intergene correlation dramatically as well (16). To some extent, the δ-sequence method can be seen as a normalization procedure because it also removes any slide-specific noise. The main difference between a classical normalization procedure and the δ-sequence method is that the latter do not attempt to replace the original variable with a distorted one. Taking the global normalization (24) as an example, after the normalization, the log-transformed expression of the *i*th gene is replaced by $x_i - \bar{x}$, where $\bar{x}$ is the sample average of log-transformed expression taken over *all genes*. As a consequence, the *i*th gene is declared to be differentially expressed if $x_i - \bar{x}$, instead of x_i is distributed differently between two phenotypes. From Table 1, we can see that with global normalization, we have fewer false positives and more true positives comparing to the not normalized data. The reason is that global normalization breaks the correlation between genes, and removes possible slide-specific noise. As a result, $x_i - \bar{x}$ is "shaper" than the original signal x_i. This performance gain comes at a price however, because $x_i - \bar{x}$ is a distorted version of x_i which does not necessarily reflect the true underlying signal. For example, consider a very simple case, all but 10% of all genes are differentially expressed. For the *i*th gene that is not differentially expressed, an artificial difference is introduced to this gene because x_i is replaced with $x_i - \bar{x}$ and $\bar{x}$ is the average of both the **DE** genes and the **NDE** genes. It results in excessive false positives even when the sample size is large.

As a comparison, δ_i is defined as the difference of only two variables. So if δ_i is distributed differently for two phenotypes, at least one of the two variables must be differentially expressed. If a given gene is declared to be **DE** for two different combinations of pairs, then it is very unlikely to be an **NDE** gene. That is the very idea of the shift method. So being a *local* variable transformation method, meaning the δs are functions of two variables instead of all or very large subset of all variables is one of its unique strength.

The only true disadvantage of the δ-sequence method is that some information is inevitably lost in the pairing of two variables. So its power is slightly lower than that of the other methods. Further research should be done to address this issue.

Acknowledgments

This research is supported by NIH Grant GM079259 (X. Qiu) and by Theodosius Dobzhansky Center for Genome Bioinformatics (L. Klebanov).

References

1. Efron B (2003) Robbins, empirical Bayes and microarrays. Ann Stat 31:366–378
2. Efron B (2004) Large-scale simultaneous hypothesis testing: the choice of a null hypothesis. J Am Stat Assoc 99:96–104
3. Efron B, Tibshrani R, Storey JD, Tusher V (2001) Empirical Bayes analysis of a microarray experiment. J Am Stat Assoc 96:1151–1160
4. Allison DB, Gadbury GL, Heo M, Fern'andez JR, Les C-K, Prolla JA, Weindruch R (2002) A mixture model approach for the analysis of microarray gene expression data. Comput Stat Data Anal 39:1–20
5. Dalmasso C, Broët P, Moreau T (2005) A simple procedure for estimating the false discovery rate. Bioinformatics 21(5):660–668
6. Pounds S, Cheng C (2004) Improving false discovery rate estimation. Bioinformatics 20:1737–1745
7. Pounds S, Morris SW (2003) Estimating the occurrence of false positives and false negatives in microarray studies by approximating and partitioning the empirical distribution of *p*-values. Bioinformatics 19:1236–1242
8. Reiner A, Yekutieli D, Benjamini Y (2003) Identifying differentially expressed genes using

false discovery rate controlling procedures. Bioinformatics 19:368–375

9. Storey JD (2002) A direct approach to false discovery rates. J R Stat Soc Ser B 64: 479–498
10. Storey JD, Tibshirani R (2003) Statistical significance for genomewide studies. Proc Natl Acad Sci USA 100(16):9440–9445
11. Tsai C-A, Hsueh H-M, Chen JJ (2003) Estimation of false discovery rates in multiple testing: application to gene microarray data. Biometrics 59:1071–1081
12. Qiu X, Klebanov L, Yakovlev A (2005) Correlation between gene expression levels and limitations of the empirical Bayes methodology for finding differentially expressed genes. Stat Appl Genet Mol Biol 4:34
13. Qiu X, Yakovlev A (2006) Some comments on instability of false discovery rate estimation. J Bioinform Comput Biol 4(5):1057–1068
14. Storey JD, Taylor JE, Siegmund D (2003) Strong control, conservative point estimation and simultaneous conservative consistency of false discovery rates: a unified approach. J R Stat Soc Ser B 66:187–205
15. Storey JD, Tibshirani R (2003) Statistical significance for genomewide studies. Proc Natl Acad Sci USA 100:9440–9445
16. Qiu X, Brooks AI, Klebanov L, Yakovlev A (2005) The effects of normalization on the correlation structure of microarray data. BMC Bioinform 6:120
17. Klebanov L, Jordan C, Yakovlev A (2006) A new type of stochastic dependence revealed in gene expression data. Stat Appl Genet Mol Biol 5 (Article7)
18. Almudevar A, Klebanov LB, Qiu X, Salzman P, Yakovlev AY (2006) Utility of correlation measures in analysis of gene expression. NeuroRx 3(3):384–395
19. Klebanov L, Yakovlev A (2006) Treating expression levels of different genes as a sample in microarray data analysis: is it worth a risk? Stat Appl Genet Mol Biol 5, Article 9
20. Qiu X, Xiao Y, Gordon A, Yakovlev A (2006) Assessing stability of gene selection in microarray data analysis. BMC Bioinform 7:50
21. Benjamini Y, Hochberg Y (2000) On the adaptive control of the false discovery rate in multiple testing with independent statistics. J Educ Behav Stat 25(1):60
22. Yeoh E-J, Ross ME, Shurtleff SA, Williams WK, Patel D, Mahfouz R, Behm FG, Raimondi SC, Relling MV, Patel A, Cheng C, Campana D, Wilkins D, Zhou X, Li J, Liu H, Pui C-H, Evans WE, Naeve C, Wong L, Downing JR (2002) Classification, subtype discovery, and prediction of outcome in pediatric acute lymphoblastic leukemia by gene expression profiling. Cancer Cell 1(2):133–143
23. Westfall PH, Young S (1993) Resampling-based multiple testing. Wiley, New York, NY
24. Yang YH, Dudoit S, Luu P, Lin DM, Peng V, Ngai J, Speed TP (2002) Normalization for cdna microarray data: a robust composite method addressing single and multiple slide systematic variation. Nucleic Acids Res 30(4):e15
25. Bolstad BM, Irizarry RA, Astrand M, Speed TP (2003) A comparison of normalization methods for high density oligonucleotide array data based on variance and bias. Bioinformatics 19:185–193
26. Gordon A, Glazko G, Qiu X, Yakovlev A (2007) Control of the mean number of false discoveries, Bonferroni, and stability of multiple testing. Ann Appl Stat 1(1):179–190

Chapter 5

Using of Normalizations for Gene Expression Analysis

Peter Bubelíny

Abstract

Normalizations of gene expression data are commonly used in practice. They are used for removing systematic variation which affects the measure of gene expression levels. But one can object to the using of normalized data for testing hypotheses. By using normalized data, tests can break nominal level of multiple testing on which we would like to test the hypotheses. It could bring a lot of false positives, which we would like to prevent. In this chapter, by simulating data with similar correlation structure as real data, we try to find out how quantile, global, and δ-sequence normalizations hold the nominal level of Bonferroni multiple testing procedure.

Key words: Gene expression, Normalization, δ-Sequence

1. Introduction

Microarray experiment produces gene expressions for a lot of genes. Consider slides (samples) of such gene expressions for two stages, i.e., slides for people with two different diagnoses. Our aim is to decide which genes are differentially expressed in these two stages. That is, we would like to test the hypotheses H_i: *the ith gene is not differentially expressed* against the alternatives A_i: *the ith gene is differentially expressed* for each gene simultaneously.

But there are some problems. One of them is that we have a lot of genes (several thousands), so we have to test a large number of hypotheses and we have only a few slides of gene expressions for both stages (several tens). Another problem of microarray data are high correlations between genes on each slide. In Fig. 1, there are histograms of 100,000 pairwise correlations of $\log_2$-expressions between randomly chosen genes from HYPERDIP and TEL data (two stages of leukemia consist of 7,084 genes and 88 and 79 slides, respectively). We can see these correlations take values close to 1 (average correlation coefficient for HYPERDIP data is 0.91

Andrei Y. Yakovlev et al. (eds.), *Statistical Methods for Microarray Data Analysis: Methods and Protocols*, Methods in Molecular Biology, vol. 972, DOI 10.1007/978-1-60327-337-4_5, © Springer Science+Business Media New York 2013

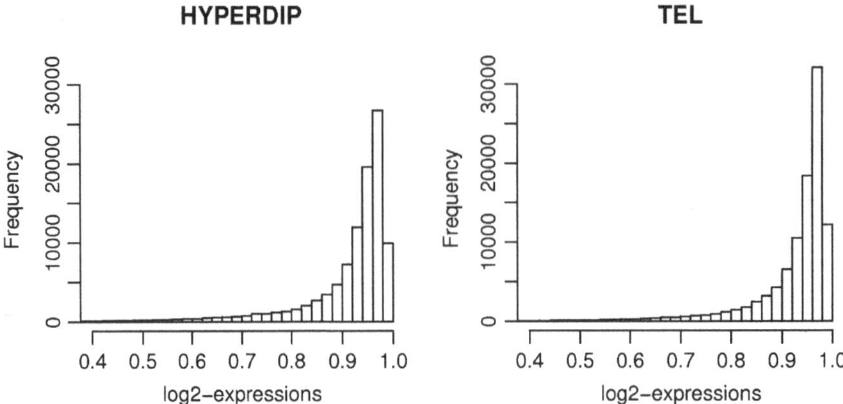

Fig. 1. Histograms of 100,000 estimates of random pairwise correlations of $\log_2$-gene expressions for HYPERDIP and for TEL data.

and 0.92 for TEL data). Many multiple testing procedures are derived under independency of hypotheses. Moreover, the correlation structure affects the power of the tests. In many chapters, a number of normalizations are considered to partially handle this problem. Testing of aforementioned hypotheses is carried out on these transformed (normalized) data. But one can object to equality of testing with nonnormalized data and with normalized data. Using normalized data, tests can break nominal level of multiple testing on which we would like to test hypotheses. It could bring a lot of false positives (number of rejected true hypotheses), which we try to prevent.

In this chapter, we use several type of normalizations on simulated data. We will test hypotheses of absence of differentially expressed genes for both for nonnormalized and normalized data. We will consider situations with different number of observations and different proportions of false hypotheses among all hypotheses. We apply Bonferroni procedure which does not only family-wise error rate (FWER) control but controls per-family error rate (PFER) as well.

2. Normalizations

There are many sources of systematic variation in microarray experiments which affect the measure of gene expression levels. The common way of removing such variations is to normalize the data. In this chapter, we compare three types of normalizations: quantile normalization, global normalization, and δ-sequence.

2.1. Quantile Normalization

The goal of the quantile normalization is to make the distribution of $\log_2$-gene expressions on each slide in a set of slides the same.

The method is motivated by the idea of a n-dimensional quantile-quantile plot. In [1] we can find the algorithm for computing X_{QN} (the matrix of $\log_2$-expressions after the quantile normalization). It is as follows.

1. Given n slides of $\log_2$-gene expressions of length m, form a matrix X of dimension $m \times n$ where each slide is a column.
2. Sort each column of X to give X_{sort}.
3. Take means across rows of X_{sort} and assign this mean to each element in the row to get X'_{sort}.
4. Obtain X_{QN} by rearranging each column of X'_{sort} to have the same ordering as the original X.

In this chapter, we use function normalize.quantiles of package *affy* (Bioconductor) of statistical software R to compute the X_{QN}.

For the two-stage problem, there are two options of how to use the quantile normalization. The first possibility is to use the quantile normalization separately for each stage. The second possibility is to create one pooled matrix of data and then to make the quantile normalization on this pooled matrix. In this chapter, we will consider both of the mentioned cases of the quantile normalization.

2.2. Global Normalization

Global normalization of gene expression levels comes from the idea that gene expression is a product of two factors. The first factor is associated with gene production and the second factor is a constant unique for each slide. Therefore, we can imagine $\log_2$-expression x_{ij} of gene i from slide j as a sum of two factors, the first factor depends on the gene i and the second depends on the slide j, this is $x_{ij} = g_i + s_j$. Hence, it seems reasonable to subtract the specific factor $S_j = s_j + c$ (c is a constant independent on i and j) from each $\log_2$-expression. There are two reasonable choices of S_j. The first is slide mean, the second is the slide median. Thus, the algorithm for computing X_{mean} (the matrix of $\log_2$-gene expressions after the global-mean normalization) and X_{med} (the matrix of $\log_2$-gene expressions after the global-median normalization) is as follows.

1. Given n slides of $\log_2$-gene expressions of length m, form a matrix X of dimension $m \times n$ where each slide is a column.
2. Take the means or medians across the columns of X to obtain slide-specific factors $\bar{X}_j$ or X_j^{n} for $j = 1, \ldots, n$.
3. For each $j = 1, \ldots, n$ subtract from jth column of X slide-specific factor $\bar{X}_j$ (or X_j^{med}) to obtain X_{mean} (or X_{med}).

2.3. δ-Sequence

In chapter [2] Klebanov and Yakovlev defined a new type of normalization. They eliminated slide effect in a different way than global normalizations do. They defined a δ-sequence normalization as differences of nonoverlapping $\log_2$-gene expressions. $\log_2$-gene expression data after δ-sequence normalization can be defined as a m^* ($m^* = m/2$, m is tacitly assumed to be even) by n dimensional matrix consisting of the random variables $\delta_{ij} = x_{2i-1,j} - x_{2i,j}$, $i = 1,\ldots,m^*$; $j = 1, \ldots, n$, where $x_{i,j}$ are $\log_2$-gene expressions for m genes from n slides. They do not specify the order in which the genes should be sorted. So in our simulation we consider two cases of ordering of the gene expression data. The first case is a random permutation of genes and the second case (monotonic) the ordering takes place according to the p-values of gene expressions (we consider a two-sample case) and in creating δ_{ij} we pair ith and $(m^* + i)$th gene for $i = 1,\ldots,m^*$. In the remaining part of the chapter it will be tacitly assumed (without loss of generality) that the genes are in random or monotonic order.

Another problem of δ-sequence is that we examine difference of two genes and so we cannot make the decision for each gene separately. Therefore, we propose a solution to handle it. Our first proposal A is not only to pair the ith and the $(m^* + i)$th gene, but to pair the ith and $(m^* + i - 1)$th gene (and the first and the last, respectively) for $i = 2,\ldots,m^*$. If the δs calculated from some special gene are rejected both times we will consider this gene differentially expressed. If the δs calculated from some special gene are not rejected or are rejected once we will consider this gene nondifferentially expressed.

Some improvement for monotonic ordering can be achieved by computing the second B δ-sequence as the difference of the ith and the $(m - i + 1)$th gene for $i = 1,\ldots,m^*$. Decisions for each gene are the same as in the previous case.

In following proposals it is tacitly assumed, that there is at most m^* false hypotheses. It seems reasonable to assume that a gene which evokes different expression of δ-sequence is the one with a lower p-value. Therefore, our third proposal C in monotonic ordering of genes is that we will consider ith $i = 1,\ldots,m^*$ gene differentially expressed, if δ_{ij} are considered differentially expressed.

Our fourth proposal D is something like a step-down modification of the third case. We will consider the ith gene for $i = 1,\ldots,m^*$ differentially expressed, if all h hypotheses for δ_{hj} are rejected for $h = 1, \ldots, i$.

Now we explore how these normalizations change the structure of pairwise correlations of $\log_2$-expressions. In Fig. 2, there are histograms of 100,000 estimates of random pairwise correlations of $\log_2$-gene expressions after global-mean, global-median, quantile, and random δ-sequence normalization of HYPERDIP and TEL data. We can see, that these histograms are different from these of correlations of nonnormalized $\log_2$-gene expressions, because they are symmetric and concentrated about zero.

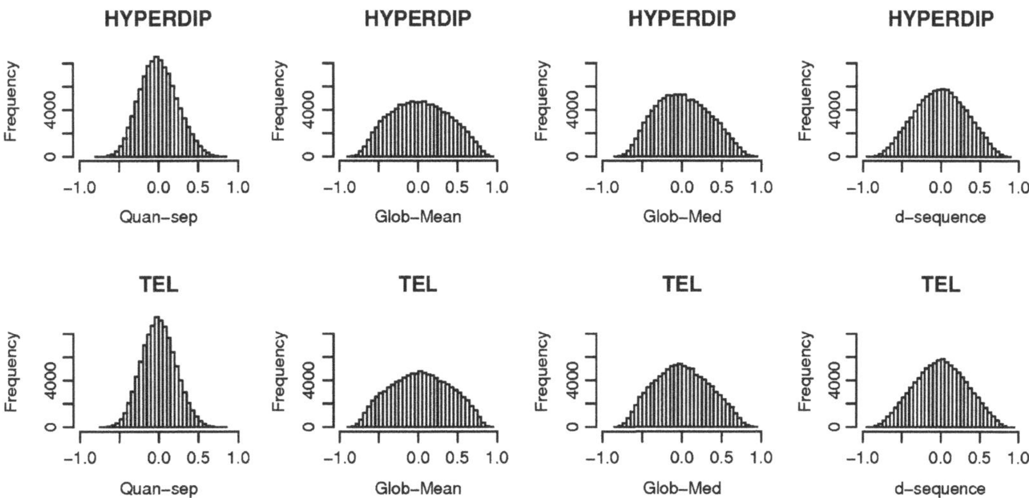

Fig. 2. Histograms of 100,000 estimates of random pairwise correlations of $\log_2$-gene expressions after normalization. From *left* to *right* is Quantile-separate, Global-Mean, Global-Median, and δ-sequence normalization. The *upper row* of histograms is for HYPERDIP data, the *lower row* is for TEL data.

3. Simulations

We know that gene expressions are highly correlated in each slide and their $\log_2$-expressions have approximately normal distribution. Therefore, we will simulate $\log_2$-gene expressions x_{ij}, $i=1, \ldots, m$ (number of genes), $j=1, \ldots, n$ (number of slides) as random variables (highly correlated in each slide) from normal distribution. We will use the following algorithm.

1. Generate independent random variables a_j and y_{ij}, $i=1, \ldots, m$; $j=1, \ldots, n$ from the standard normal distribution.
2. For a fixed ρ, define $\log_2$-expressions as $x_{ij} = \sqrt{\rho^0 a_j} + \sqrt{1-\rho^0}\, y_{ij}$, $i=1, \ldots, m$; $j=1, \ldots, n$.

These simulated $\log_2$-expressions have central normal distribution and moreover for each $i_1 \neq i_2$ they have all pairwise correlations $\operatorname{corr}(x_{i_1 j}, x_{i_2 j}) = \rho$ and for each i, $k=1, \ldots, m$ and $j \neq l$ we have $\operatorname{corr}(x_{ij}, x_{k,l}) = 0$.

Throughout the chapter, we take $\rho = 0.9$ and the number of genes equal to 500 (we consider only 500 genes because of computational complexity). We will generate $\log_2$-expressions for two stage by aforementioned algorithm. We consider cases with equal number of slides for each stage $n = n_1 = n_2 = 10, 25 \text{ and } 50$. In the second stage, we modify $k=24$, 90, and 200 genes (we create k false hypotheses). We will consider three different alternatives. In the first alternative we add a constant C, $C=0.8$ for $n=10$; $C=0.44$ for $n=25$ and $C=0.3$ for $n=50$ to k $\log_2$-expressions on each slide of the second stage. In the second alternative, we shift k $\log_2$-gene expressions on each slide of the second stage so that their expectations create a random vector $\mu = (\mu_1, \ldots, \mu_k)$ with i.i.d. components from standard normal distribution. The third alternative is a combination of the previous two alternatives. So, expected values of k genes for the second stage of simulated $\log_2$-expressions form random vector with i.i.d. components from $N(C,1)$ $C=0.8$ for $n=10$; $C=0.44$ for $n=25$ and $C=0.3$ for $n=50$.

For each of 27 cases, we run 5,000 simulations. For each simulated dataset we calculate quantile-separate, quantile-pooled, global-mean, global-median, and proposed δ-sequence normalizations. We will study average number of true positive (measure of power), mean number of false positive (estimate of PFER) and estimate family-wise error rate $\equiv P(FP>0)$ (as relative frequency of present false positives among rejected hypotheses) according to Bonferroni method at nominal level $\alpha=0.05$ by Welsh t-test and N-test. That means, that the hypothesis is rejected if its unadjusted p-value is lower than 0.05/500 (for some δ-sequences lower than 0.5/250).

4. Results

In all situations Welsh t-test and N-test give very similar results. In the most of the cases Welsh t-test seems a little more powerful, but the difference is very small. Therefore, it is overbold to say that the t-test is better than the N-test. Consequently, we do not give results of simulation for the N-test in this chapter.

The results of simulations for the t-test for alternative one are given in Table 1. In Table 2 there are results for alternative two. Finally, the results for alternative three are in Table 3. From these tables we can see that the results for expressions without normalization (*none*) hold nominal level for FWER and PFER, but the power is very low. If we look at separately quantile normalized data (*Q-sep*), we can see that this normalization produces too many false positives. The quantile normalization for pooled data (*Q-pool*) works much better. But still for alternative one and alternative three there are too many false positives. Global-mean (*gl-mean*) and global-median (*gl-med*) normalizations produce similar amount of false positives as the quantile normalization for pooled data. Among global normalizations, global median is better than global-mean, but there is still too much false positives in cases with large number of false hypotheses. Therefore, the global normalization (and quantile normalizations as well) is not suitable to make decision about which gene is differentially expressed and which is not. On the other hand, if we look at the results for δ-sequence (*D-rand* and *D-mono*) we can see that it holds the nominal level. However, the δ-sequence

Table 1

Results of simulation for the alternative one

	$k=24$			$k=90$			$k=200$		
Norm.	$\overline{TP}$	$\overline{FP}$	$\hat{\alpha}$	$\overline{TP}$	$\overline{FP}$	$\hat{\alpha}$	$\overline{TP}$	$\overline{FP}$	$\hat{\alpha}$
n=10									
None	0.154	0.015	0.003	0.459	0.022	0.001	1.215	0.005	0.001
Q-sep	13.964	59.708	0.708	52.632	51.173	0.690	114.198	36.848	0.650
Q-pool	12.057	0.040	0.039	17.183	0.090	0.085	5.920	1.041	0.647
Gl-mean	14.65	0.056	0.056	34.796	0.375	0.311	21.391	4.983	0.995
Gl-med	15.474	0.042	0.041	45.723	0.114	0.108	33.405	2.325	0.901
D-rand	6.668	0.034	0.032	21.863	0.026	0.025	35.476	0.017	0.017
D-mono	6.328	0.03	0.03	23.536	0.031	0.031	49.213	0.031	0.031
D-rand-A	2.504	2.489	0.917	8.551	8.62	1	14.377	14.479	1
D-mono-A	3.622	1.754	0.695	13.872	7.06	0.983	30.007	14.396	0.997
D-mono-B	2.494	0.013	0.012	8.382	0.203	0.098	13.371	3.544	0.85
D-mono-C	5.234	0.481	0.128	19.278	2.332	0.178	41.025	5.291	0.223
D-mono-D	1.166	0.005	0.004	1.768	0.019	0.017	1.828	0.06	0.043
n=25									
None	0.122	0.0502	0.002	0.586	0.025	0.003	1.282	0.024	0.002
Q-sep	14.006	92.230	0.755	53.322	76.327	0.746	115.680	55.632	0.719
Q-pool	15.144	0.060	0.059	30.697	0.281	0.246	14.808	3.2586	0.967
Gl-mean	15.824	0.058	0.056	38.349	0.406	0.331	23.911	5.534	0.996
Gl-med	16.212	0.054	0.052	42.857	0.264	0.23	27.002	4.542	0.989
D-rand	7.064	0.04	0.039	22.799	0.03	0.029	37.031	0.016	0.016
D-mono	6.408	0.049	0.049	22.903	0.036	0.036	43.929	0.026	0.025
D-rand-A	2.839	2.85	0.952	9.579	9.634	1	16.237	16.372	1
D-mono-A	3.734	2.429	0.818	13.749	9.194	0.994	27.328	16.898	0.986
D-mono-B	3.09	0.021	0.017	10.275	0.302	0.105	13.614	2.389	0.547
D-mono-C	5.029	0.595	0.144	17.982	2.532	0.189	34.96	5.868	0.233
D-mono-D	1.942	0.007	0.007	3.541	0.025	0.022	2.685	0.057	0.042
n=50									
None	0.19	0.083	0.004	0.616	0.026	0.002	1.352	0.048	0.002
Q-sep	14.096	94.089	0.762	53.571	81.440	0.748	118.078	61.575	0.730
Q-pool	15.785	0.066	0.064	35.519	0.376	0.312	19.760	4.525	0.992
Gl-mean	16.17	0.062	0.06	39.615	0.424	0.343	24.875	5.862	0.998
Gl-med	16.361	0.057	0.055	41.784	0.339	0.286	26.233	5.36	0.996
D-rand	7.122	0.042	0.041	23.192	0.03	0.029	37.694	0.017	0.017
D-mono	6.456	0.047	0.047	22.528	0.04	0.04	40.808	0.027	0.026
D-rand-A	2.926	2.912	0.954	9.981	9.949	1	16.925	16.875	1
D-mono-A	3.862	2.699	0.861	13.797	10.203	0.998	25.502	17.998	0.985
D-mono-B	3.167	0.026	0.021	10.374	0.304	0.095	12.457	2.134	0.384
D-mono-C	5.093	0.615	0.151	17.583	2.643	0.199	31.268	6.818	0.253
D-mono-D	2.468	0.007	0.007	4.871	0.025	0.021	3.206	0.067	0.048

Number of genes $N=500$; number of slides in each stage $n=10, 25, 50$; number of false hypotheses $k=24, 90, 200$; $\overline{TP}$ —average of true positives, $\overline{FP}$ —average of false positives and $\hat{\alpha}$ —estimate of FWER

Table 2

Results of simulation for the alternative two

Norm.	k=24 $\overline{TP}$	$\overline{FP}$	$\hat{\alpha}$	k=90 $\overline{TP}$	$\overline{FP}$	$\hat{\alpha}$	k=200 $\overline{TP}$	$\overline{FP}$	$\hat{\alpha}$
n=10									
None	1.279	0.015	0.003	4.789	0.018	0.003	10.331	0.011	0.003
Q-sep	12.472	57.945	0.702	46.579	47.668	0.693	103.639	35.858	0.665
Q-pool	10.626	0.040	0.039	31.016	0.025	0.025	45.536	0.013	0.013
Gl-mean	11.602	0.036	0.035	43.494	0.034	0.033	96.741	0.032	0.032
Gl-med	11.605	0.033	0.032	43.456	0.031	0.031	96.616	0.03	0.029
D-rand	8.6	0.032	0.032	30.934	0.023	0.023	63.983	0.015	0.014
D-mono	8.568	0.05	0.049	31.983	0.05	0.05	70.588	0.044	0.043
D-rand-A	4.222	3.87	0.985	17.905	13.558	1	48.227	24.487	1
D-mono-A	7.45	5.194	0.99	28.628	20.135	1	68.355	41.144	1
D-mono-B	6.804	0.01	0.008	25.61	0.126	0.093	57.778	0.827	0.363
D-mono-C	7.979	0.639	0.287	29.814	2.219	0.432	66.449	4.182	0.447
D-mono-D	5.774	0	0	17.388	0	0	33.38	0	0
n=25									
None	6.024	0.087	0.002	22.566	0.079	0.003	50.454	0.014	0.002
Q-sep	17.130	85.632	0.748	64.369	76.407	0.756	143.233	58.376	0.722
Q-pool	16.468	0.039	0.038	56.568	0.030	0.030	106.756	0.020	0.019
Gl-mean	16.924	0.052	0.051	63.436	0.056	0.054	141.13	0.055	0.052
Gl-med	16.916	0.047	0.046	63.407	0.044	0.044	141.015	0.037	0.037
D-rand	14.46	0.043	0.042	51.074	0.034	0.033	102.575	0.014	0.014
D-mono	14.637	0.079	0.078	54.751	0.089	0.089	121.124	0.086	0.085
D-rand-A	7.643	7.155	1	33.057	26.246	1	89.471	49.943	1
D-mono-A	13.767	11.724	1	52.701	44.909	1	124.308	94.741	1
D-mono-B	13.611	0.024	0.018	51.364	0.35	0.194	116.134	7.116	0.9
D-mono-C	13.912	0.803	0.336	52.004	2.836	0.443	115.762	5.448	0.467
D-mono-D	12.159	0.001	0.001	43.065	0	0	93.567	0	0
n=50									
None	10.299	0.02	0.002	38.504	0.052	0.002	85.72	0.004	0.002
Q-sep	19.234	96.293	0.766	72.112	81.258	0.761	160.453	63.197	0.729
Q-pool	18.822	0.048	0.046	67.018	0.034	0.033	136.251	0.032	0.031
Gl-mean	19.133	0.056	0.055	71.794	0.077	0.071	159.653	0.103	0.09
Gl-med	19.133	0.049	0.049	71.765	0.044	0.042	159.61	0.038	0.038
D-rand	17.164	0.046	0.046	60.546	0.03	0.03	120.118	0.019	0.019
D-mono	17.453	0.086	0.086	65.524	0.088	0.088	145.238	0.112	0.112
D-rand-A	9.224	8.757	1	40.788	32.738	1	110.474	63.557	1
D-mono-A	16.841	15.062	1	64.193	57.732	1	149.796	124.055	1
D-mono-B	16.754	0.018	0.015	63.531	0.427	0.22	144.2	30.473	1
D-mono-C	16.866	0.673	0.303	63.102	2.51	0.44	140.59	4.76	0.449
D-mono-D	15.626	0.003	0.003	56.892	0	0	126.965	0	0

Number of genes N=500; number of slides in each stage n=10, 25, 50; number of false hypotheses k=24, 90, 200; $\overline{TP}$—average of true positives, $\overline{FP}$—average of false positives and $\hat{\alpha}$—estimate of FWER

Table 3

Results of simulation for the alternative three

Norm.	$k=24$			$k=90$			$k=200$		
	$\overline{TP}$	$\overline{FP}$	$\hat{\alpha}$	$\overline{TP}$	$\overline{FP}$	$\hat{\alpha}$	$\overline{TP}$	$\overline{FP}$	$\hat{\alpha}$
n=10									
None	2.72	0.048	0.003	10.024	0.011	0.003	21.468	0.025	0.004
Q-sep	14.951	59.709	0.706	55.806	51.985	0.694	123.363	36.617	0.661
Q-pool	12.812	0.036	0.035	33.883	0.019	0.019	43.448	0.042	0.042
Gl-mean	14.299	0.056	0.055	51.043	0.414	0.339	106.236	5.225	0.983
Gl-med	14.457	0.04	0.039	53.135	0.059	0.056	114.076	0.213	0.191
D-rand	11.62	0.039	0.038	40.448	0.026	0.025	79.657	0.017	0.017
D-mono	11.76	0.051	0.05	43.705	0.041	0.041	95.92	0.055	0.055
D-rand-A	5.845	5.533	0.999	24.063	19.545	1	61.064	35.844	1
D-mono-A	10.701	8.467	1	40.735	32.414	1	94.855	66.891	1
D-mono-B	10.006	0.01	0.007	37.165	0.185	0.115	82.294	1.808	0.529
D-mono-C	11.185	0.626	0.245	41.675	2.071	0.385	91.386	4.59	0.434
D-mono-D	8.976	0	0	28.546	0	0	55.563	0	0
n=25									
None	7.076	0.013	0.003	26.371	0.026	0.002	58.539	0.009	0.003
Q-sep	17.720	87.353	0.757	66.466	74.890	0.740	147.914	54.586	0.722
Q-pool	16.962	0.041	0.040	57.486	0.052	0.050	105.681	0.139	0.130
Gl-mean	17.543	0.071	0.067	65.102	0.496	0.367	143.112	6.897	0.957
Gl-med	17.567	0.045	0.044	65.558	0.065	0.063	144.731	0.266	0.232
D-rand	15.233	0.049	0.047	53.528	0.033	0.032	106.361	0.016	0.016
D-mono	15.468	0.077	0.076	57.801	0.076	0.076	127.777	0.092	0.092
D-rand-A	8.048	7.642	0.999	34.95	27.942	1	93.181	53.295	1
D-mono-A	14.671	12.711	1	55.926	48.411	1	131.073	102.757	1
D-mono-B	14.494	0.021	0.016	54.606	0.35	0.191	123.232	10.679	0.967
D-mono-C	14.789	0.756	0.314	55.176	2.701	0.441	122.431	5.438	0.465
D-mono-D	13.185	0.001	0.001	46.694	0	0	101.784	0	0
n=50									
None	10.75	0.026	0.002	40.241	0.067	0.004	89.702	0.047	0.003
Q-sep	19.505	100.495	0.773	72.847	79.662	0.754	162.091	60.497	0.734
Q-pool	19.028	0.051	0.050	67.474	0.083	0.078	136.153	0.329	0.253
Gl-mean	19.354	0.078	0.075	72.352	0.627	0.382	160.207	8.519	0.923
Gl-med	19.358	0.049	0.048	72.445	0.072	0.068	160.712	0.326	0.263
D-rand	17.466	0.049	0.047	61.329	0.033	0.033	121.454	0.018	0.018
D-mono	17.736	0.085	0.085	66.511	0.103	0.102	147.416	0.113	0.113
D-rand-A	9.422	8.932	1	41.414	33.392	1	111.946	64.634	1
D-mono-A	17.141	15.418	1	65.288	59.046	1	151.981	126.755	1
D-mono-B	17.088	0.023	0.016	64.612	0.418	0.212	146.579	33.914	1
D-mono-C	17.144	0.676	0.309	64.159	2.455	0.433	142.775	4.754	0.466
D-mono-D	15.983	0.002	0.002	58.297	0	0	129.563	0	0

Number of genes $N=500$; number of slides in each stage $n=10, 25, 50$; number of false hypotheses $k=24, 90, 200$; $\overline{TP}$ —average of true positives, $\overline{FP}$ —average of false positives and $\hat{\alpha}$ —estimate of FWER does not give decisions for each gene, but only for pairs of genes. Therefore, the δ-sequence seems to be a suitable normalization to decide which genes are differentially expressed and which are not. The problem with the δ-sequence is how to make a decision for each gene separately. Now we look at our proposals to handle it.

The first proposal pairs the ith gene with the $(m+i)$th and $(m^* + i - 1)$th gene, so we construct two δ-sequences. In the tables, this proposal is *D-mono-A*. But we can see that this is not a good solution. It produces too many false positives. It can be caused by frequently pairing nondifferentially expressed genes with two differentially expressed genes. The second proposal (*D-mono-B*) for monotonically ordered genes gives some improvement, but still it is not as good as we need. The third proposal (*D-mono-C*) is better than the proposal A, but it is worse than the second proposal. So, it still does not work as well as we would like to. Finally, the fourth proposal (*D-mono-D*) holds the nominal level in all simulation cases, even in a lot of cases it undervalues the nominal level too much. Moreover, in all simulated cases it has a better power than nonnormalized data. Therefore, this proposal seems to be a good improvement for two-sample testing of differentially expressed genes.

5. HYPERDIP and TEL Data

Now we work with HYPERDIP and TEL data for childhood leukemia consisting of 7,084 genes and from 88 and 79 slides, nonrespectively. For these data, we test which genes are differentially expressed. We would like to compare nonnormalized testing with our best behaved fourth proposal D. If we apply classical approach with the t-test for nonnormalized data, we will find 72 differentially expressed genes. For the N-test this number is 73 genes (67 genes are the same for the N-test and the t-test). If we test according to proposal D, we will have 87 differentially expressed genes by the t-test and 93 differentially expressed genes by the N-test (81 genes are the same). These results confirm that proposal D is an improvement of the classical approach using normalized data.

6. Conclusion

In this chapter we studied the quality of testing of simulated microarray data after normalizations which are used in practical applications for microarray data and found some improvement of the classical approach working with nonnormalized data. We simulated data with similar correlation structure as real data. We normalized these simulated data and we tested which genes are differentially expressed. We looked at how many true positives and how many and how often false positives are produced by testing normalized data. From these three measures we tried to decide which considered normalizations we should use and which of them is the best. We studied cases with different number of slides and with different proportion of false hypotheses for three different alternatives. We used Bonferroni multiple testing procedure to derive the adjusted p-values. The raw p-values were computed by the Welsh t-test and by the N-test.

In our simulation study the N-test works very similarly to the t-test. Simulation study showed that testing with nonnormalized data produces much less true positives than testing with normalized data. Moreover, we found out that quantile normalizations and global normalizations produce too many false positives in many situations. Therefore, we do not recommend using these normalizations to decide which genes are differentially expressed. We designed some procedures for deciding which gene is differentially expressed using δ-sequences. Only one of our proposals D holds the nominal level of multiple testing for all simulated cases. Moreover, it has a better power in simulation and for real data than the approach using gene expressions without any normalization. But our proposal undervalues the nominal level in a lot of cases. Therefore, there is still some space to improve our procedure.

Acknowledgment

Author thanks Prof. Lev Klebanov for valuable comments, remarks, and overall help.

References

1. Bolstad M, Irizarry R, Strand M, Speed T (2003) A comparison of normalization methods for high density oligonucleotide array data based on variance and bias. Bioinformatics 19(2):185–193
2. Klebanov L, Yakovlev A (2007) Diverse correlation structures in gene expression data and their utility in improving statistical inference. Ann Appl Stat 1(2):538–559

Chapter 6

Constructing Multivariate Prognostic Gene Signatures with Censored Survival Data

Derick R. Peterson

Abstract

Modern high-throughput technologies allow us to simultaneously measure the expressions of a huge number of candidate predictors, some of which are likely to be associated with survival. One difficult task is to search among an enormous number of potential predictors and to correctly identify most of the important ones, without mistakenly identifying too many spurious associations. Mere variable selection is insufficient, however, for the information from the multiple predictors must be intelligently combined and calibrated to form the final composite predictor. Many commonly used procedures overfit the training data, miss many important predictors, or both. Although it is impossible to simultaneously adjust for a huge number of predictors in an unconstrained way, we propose a method that offers a middle ground where some partial multivariate adjustments can be made in an adaptive fashion, regardless of the number of candidate predictors. We demonstrate the performance of our proposed procedure in a simulation study within the Cox proportional hazards regression framework, and we apply our new method to a publicly available data set to construct a novel prognostic gene signature for breast cancer survival.

Key words: Micro-array data, Prognostic signature, Gene selection, Shrinkage estimation, Cox proportional hazards regression, Censored survival data

1. Introduction

Modern micro-array technologies now allow us to simultaneously measure the expression of a huge number of different genes and proteins, some of which are likely to be associated with cancer prognosis. While such gene expressions are unlikely to ever replace important clinical covariates, evidence is already mounting that they can provide a significant amount of additional predictive information (1–10). The difficult task, then, is to search among an enormous number of potential predictors and to correctly identify most of the important ones, without mistakenly identifying too many spurious associations.

Andrei Y. Yakovlev et al. (eds.), *Statistical Methods for Microarray Data Analysis: Methods and Protocols*, Methods in Molecular Biology, vol. 972, DOI 10.1007/978-1-60327-337-4_6,

Many commonly used procedures unfortunately overfit the training data, leading to subsets of selected genes many of which are unrelated to the outcome in the target population, despite appearing predictive in the particular sample of data used for subset selection. Part of the reason for this lies in the extreme biases induced by applying standard estimation methods as if genes had not been picked based on their estimates being larger than those of other genes. Thus, our proposed method tackles not only the issue of identifying important genes but also that of better estimating their relationships with the outcome. Indeed, such improved estimation is a key component to selecting appropriate genes in the first place, since selection is typically based upon estimates. Both gene selection and parameter estimation are critical components in constructing valid prognostic gene signatures, and these tasks should ideally be intertwined.

Some genes might only be useful predictors when used in concert with certain other related genes and/or with clinical covariates, yet the vast majority of available methods are inherently univariate in nature, based only on the marginal associations between each predictor and the outcome. While it is impossible to simultaneously adjust for a huge number of predictors in an unconstrained way, we aim to provide a middle ground where some partial adjustments can be made, even in the face of a huge number of candidate predictors.

One multivariate approach that has been considered by some is to use principal components analysis (PCA) to construct $n-1$ or fewer orthogonal *eigengenes*, each of which is a linear combination of all p (typically log-transformed) gene expressions [HTF2001]. Associations between these eigengenes and the clinical outcome can then be explored in hopes of constructing prognostic signatures (10). However, unless some (typically *ad hoc*) approach is implemented to exclude the contributions of the majority of the genes from any given eigengene, no gene selection is performed and the predictive signal in the truly important genes is typically swamped by the noise of the many others. On the other hand, if many elements of the PCA loadings are thresholded to zero after PCA, the key properties of orthogonality and maximal variation among the first few eigengenes are destroyed, begging the question of why PCA was applied in the first place.

A somewhat related strategy uses the idea of partial least squares (PLS, (11–13)) to derive predictors as linear combinations of the log gene expressions. In its pure form, each PLS predictor gives weight to every gene, like PCA. However, the key difference is that these weights are a function not only of the gene expression matrix but also of the outcome. In fact, the first PLS direction is formed by performing univariate regressions of the outcome on each of the p genes and using the resulting unadjusted regression coefficients to form the first linear combination. Thus, genes with stronger marginal associations with the outcome are seemingly appropriately given more weight than are genes that are marginally unrelated to the response. The result is that models based on just a handful of PLS components fit the training data far better than models based on a large number of PCA eigengenes. In fact, PLS fits the training data far *too* well. Indeed, when p is much larger than n, even the very *first* PLS predictor can overfit the training data in the sense that (1) the magnitude of the estimated coefficients are far too large, on average, (2) the likelihood is increased by more than if all of the truly important genes were included by an oracle, and (3) the predictive performance with independent test data is miserable. Moreover, adding additional PLS components completely saturates the model in short order, long before $n-1$ (or even $\sqrt{n}$) components can be entered. The number of PLS components is not even close to equivalent to the classical notion of degrees of freedom, given that each component is formed by estimating p parameters. Variations on PLS, where many of the gene loadings are shrunken to zero via some *ad hoc* gene selection procedure might perform somewhat better (3, 4, 14). However, in the extreme case where all but one of the gene contributions is shrunken to zero for each PLS predictor, the method becomes very similar to classical forward stepwise selection. And since the stepwise selection approach is already too greedy an algorithm, in the sense that it can be shown to badly overfit the training data, the utility of these variations on PLS remains dubious.

2. Stagewise Forward Search with Shrinkage Estimation

Our approach is a stagewise forward search algorithm that uses constrained estimation, or "shrinkage," at every stage. The first step of this algorithm selects precisely the same 1st gene as would both stepwise forward selection or most any univariate approach, i.e., that gene whose contribution to the unconstrained likelihood for the univariate regression is largest. However, the parameter estimate for this gene is then iteratively shrunken toward the null value (which is typically 0) and then temporarily *fixed* at this value at the next stage where the set of candidate genes is again searched for the best gene to add to the model. At each stage, only a single parameter is allowed to freely vary in the likelihood, which helps to alleviate the problems experienced by standard stepwise selection procedures as the number of selected variables grows. Thus, this method has much in common with stagewise forward search algorithms (13) that have been used in machine learning to approximate Tibshirani's L_1-penalized least squares *lasso* (15) procedure. However, seemingly subtle differences in such algorithms can result in substantial differences in performance. If the lasso stopping rule is weak, then a large number of unrelated genes are typically selected, while a strict stopping rule results in severe underestimation of the associations between each gene and the outcome (i.e., over-shrinkage or underfitting), in addition to missing several important genes. In either case, the estimated prognostic gene signature resulting from the lasso-type approach will not perform particularly well.

Our proposed method, which we call the *shrinkstage* method, overcomes some of the difficulties associated with the lasso approach due to our novel, easily implemented estimation method used at each stage. First, rather than constraining the estimate at each stage to be below some small increment threshold as with the stagewise approximation to the lasso, we simply shrink the maximum likelihood estimate at each stage by a given shrinkage factor $0 < \gamma \leq 1$. Next, once each new variable has entered the model, we update all of the parameter estimates in the model, but in a constrained fashion. First, we iteratively further shrink any of the estimates if the reduction in the likelihood compared with unconstrained estimation (of the *single* parameter) falls below a given threshold, implied by an approximate p-value-to-shrink. Once the iterative shrinking stage is complete, we allow for an iterative growth stage, governed by an approximate p-value-to-grow. In particular, we allow each parameter estimate to grow by γ times the difference between the current estimate and the maximum likelihood estimate, provided the increase in the likelihood is sufficiently large. Once the growth stage is complete, we consider adding the next variable, according to an approximate p-value-to-enter, which implies the stopping rule. The result is a method which more fully adjusts for the effects of strong predictors early on than does the forward stagewise approximate lasso method, resulting in not only superior parameter estimates but also better selection of relevant genes. However, the parameter estimates still never grow all the way to the maximum likelihood estimates, which protects against overfitting, in contrast to stepwise subset selection methods.

3. Simulation Results

We tested the new method proposed in Subheading 2 by applying it in the critical simulation setting where we know the true generating model. In this section we describe our simulation model, the details of each of the methods we compared, the criteria we used to assess the performance of each method, and the results of our simulation experiment.

We generated 20 independent simulated sets of censored survival data, each containing $n = 125$ independent subjects and $p = 1{,}000$ correlated gene expression levels, with only 12 of the genes associated with survival. Each algorithm was applied to precisely the same 20 simulated data sets. The distribution of the log gene expression levels was multivariate normal with unit variances and an autoregressive correlation structure with maximal correlation $r = 0.5$ between any two adjacent genes. Thus, every gene was positively correlated with every other gene, yet each gene was only strongly correlated with a handful of other genes since the correlations tail off geometrically. The distribution of the *i*th patient's survival time T_i was taken to be exponential with rate $x_i^T\beta$, where x_i denotes the *p*-vector of log gene expression levels for subject *i* and β denotes the *p*-vector of log-hazards, or model parameters. The distribution of the independent random censoring time *C* was standard exponential (rate 1), which induced approximately 50% censoring in each data set since the observed survival data for each subject is the follow-up time $Y_i = \min(T_i, C_i)$ and the censoring indicator $\Delta_i = I(T_i \le C_i)$. For simplicity, each element of the parameter vector β contained one of three possible values: 0 (unrelated gene), 1 (up-regulation increases risk by 2.718-fold per SD), or −1 (down-regulation increases risk by 2.718-fold per SD). The magnitude of 1 was chosen so as to be large enough to be detectable by good procedures yet not so huge as to be easily detectable by even poor procedures. Rather than vary the magnitudes of the coefficients in this particular experiment, we chose to vary only the correlations between the relevant genes, as follows. Out of the 1,000 autocorrelated genes, 988 were unrelated to survival, while the following 12 were relevant predictors: 150, 151, 300, 302, 450, 453, 600, 604, 750, 755, 900, 906; and the corresponding 12 nonzero coefficients were 1, 1, −1, −1, 1, 1, −1, −1, 1, 1, −1, −1; thus, risk was associated with up-regulation for six genes and down-regulation for another six. The correlation between log gene expressions 150 and 151 was 0.5, that between 300 and 302 was 0.25, and so on, so that the correlation for genes 900 and 906 was only 0.015625. Also note that, e.g., irrelevant gene 301 had correlation 0.5 with both genes 300 and 302 and noise gene 602 has correlation 0.25 with both genes 600 and 604.

The following performance criteria were computed for each method applied to each of the 20 simulated data sets. We counted the number of relevant genes identified (out of 12), the number of irrelevant genes additionally selected (out of 988), and the total number of genes selected (or the size of each model). Since the goal here is not just to identify genes but to construct a complete prognostic gene profile, i.e., a linear combination of log gene expression levels that predicts survival, good estimation of the log hazard ratios for each gene is also critical. Thus, as our primary criterion to judge the predictive performance of each method, we used a version of the Mean Squared Prediction Error (MSPE) for a mean zero linear predictor in the Cox Proportional hazards model, given by the multivariate distance

$$\mathrm{MSPE} = (\hat{\beta} - \beta)^T \Sigma \, (\hat{\beta} - \beta) \tag{1}$$

between the true vector β and the estimated p-vector $\hat{\beta}$, where the latter contains zeroes for every gene that was not selected, and where Σ denotes the covariance matrix of the log gene expressions. The more common definition of MSPE would be to replace Σ in Eq. 1 by its empirical random counterpart $\mathbf{X}^T\mathbf{X}/n$, where $\mathbf{X}$ denotes the $n \times p$ log gene expression matrix from an independent *test* sample. While we certainly could have done this, we note that by replacing $\mathbf{X}^T\mathbf{X}/n$ by its expectation $\Sigma + \mu_x^T \mu_x \,/\, n = \Sigma$ (since each random vector of log gene expressions has mean $\mu_x = 0$) we reduce the simulation-to-simulation variability of MSPE, while retaining all of its attractive properties, including the fact that it depends on the observed gene expression values in the *training* data only through the parameter estimates $\hat{\beta}$. The MSPE does ignore estimation of the nonparametric baseline hazard function, which is treated as an infinite-dimensional nuisance parameter when fitting the Cox model using the partial likelihood. But the MSPE takes proper account of the covariance Σ of the log gene expression values and penalizes for selecting inappropriate genes, missing relevant genes, and for estimation error (overestimating or underestimating each gene's effect). If every parameter were perfectly estimated, MSPE would be 0, so methods with the smallest MSPE provide superior predictions to those with larger MSPE.

As a weaker performance criterion, we also computed a similar analog to the correlation between the linear predictor $\hat{\beta}^T \mathbf{x}$ based on the estimated coefficients $\hat{\beta}$ and the linear predictor $\beta^T \mathbf{x}$ based on the true coefficients β, where $\mathbf{x}$ denotes a p-vector of log gene expressions from an independent *test* subject. Again, rather than actually generating a random test sample and compute an empirical correlation, we chose to derive and use the more stable formula based on inner products with respect to the true covariance matrix Σ of the log gene expressions:

$$\mathrm{cor}_{lp} = \frac{\mathrm{cov}}{\sigma(\beta)\sigma(\hat{\beta})}, \quad \text{where } \mathrm{cov} = \beta^T \Sigma \, \hat{\beta}, \sigma^2(\beta) = \beta^T \Sigma \, \beta \text{ and } \sigma^2(\hat{\beta}) = \hat{\beta}^T \Sigma \, \hat{\beta}. \tag{2}$$

This metric ranges over (−1,1), with 0 corresponding with no relationship between the estimated risk score and the true risk, 1 corresponding with a perfect linear relationship, and −1 corresponding with a perfect linear relationship in the incorrect direction. Unlike MSPE, a risk model can sometimes achieve a high cor_{lp} even when serious underfitting or overfitting results in coefficient estimates that are 100 times too small or too large, or if the estimated risk score is shifted far from the center of the true risk score. Although this is clearly a weaker performance metric than MSPE, a risk score with high cor_{lp} can still be useful for risk ranking sets of patients, and it can potentially be recalibrated to also have low MSPE, using additional data.

For the shrinkstage method, the following ad hoc control parameters were used: the shrinkage factor γ was 0.85, p-to-enter was 0.015, p-to-shrink was 0.8, and p-to-grow was 0.5. The stepwise selection algorithm was started from the empty model and expanded at each iteration with a p-to-enter of 0.002; after each variable was entered, the stepwise algorithm was also programmed to drop any previously entered variable with a p-to-drop of 0.002 so that only those variables whose Likelihood Ratio Test(LRT)-based p-values of 0.002 or smaller were retained. These nominal p-values for the stepwise algorithm were chosen so as to make the typical size of the resulting model similar to the other procedures (and close to the true operating model size of 12), and to prevent serious overfitting (which already began to occur at p-to-enter of 0.0025).

In addition to the shrinkstage and stepwise methods, the following methods were compared. To provide a guideline as to the worst tolerable MSPE, MSPE was calculated for the *empty* model (selecting no genes); any method doing worse than this (MSPE = 14), e.g., when too many unrelated genes are selected, is obviously a seriously flawed procedure. To provide an envelope as to nearly the best (perhaps unattainable) performance one might expect, the *oracle* model was fit, including the 12 relevant genes and no others. In order to assess the performance of typical univariate screening procedures, the following univariate selection procedure was compared: each parameter was estimated via univariate Cox regression, and only those genes with nominal LRT p-values less than 0.005 were selected; all of their parameters were then simultaneously estimated by including just those selected genes in a single multivariate Cox model. Again, the nominal alpha level of 0.005 was chosen so as to make the average size of the selected subset close to 12, like the other methods; a lower threshold would choose fewer noise variables but also identify fewer correct genes; we also later reran this procedure using a weaker 0.01 level, for comparison. Finally, we programmed a custom stagewise approximation to the lasso procedure in the Cox model framework, with a default novel stopping rule based on a nominal level likelihood ratio test to enter a variable or increase the coefficient on a previously entered variable by up to 2% of its range. We initially tried stopping the stagewise lasso at the nominal 0.05 level. This initial stopping rule for the stagewise lasso procedure was chosen since it was weak enough to allow most of the correct variables to enter yet stringent enough that the false discovery rate would not exceed about 50%. However, when we observed the poor MSPE of this approach in our simulation setting, we allowed the lasso to run further, checking its performance at the 0.10, 0.15, and 0.20 nominal levels. Running the lasso much further would likely become troublesome, due to the limited number of available uncensored observations (49 in one data set). We found the MSPE continued to improve as we loosened the lasso constraint, with the coefficients on the truly predictive genes growing closer to their true values, despite the large numbers of induced false discoveries (mostly with relatively small estimated coefficients). The cor_{lp}, however, remained fairly constant throughout this relatively wide range of lasso stopping points, reflecting the fact that the useful genes that lasso could find typically entered within the first 13–19 nonzero coefficients.

The medians over the 20 simulated data sets for all of the methods appear in Table 1, with the exception of the middle two lasso stopping points. With respect to MSPE, there is little to choose between univariate screening at the 0.005 level, stepwise selection at the 0.002 level, and the stagewise lasso at the 0.05 level, though the lasso is the clear winner among these three in terms of cor_{lp}, suggesting that it could be useful for risk ranking or recalibration. Relaxing the alpha level for univariate screening to 0.01 slightly improved its performance, at the expense of doubling the number of false discoveries just to find one more true predictor. Further relaxing the alpha level for univariate screening to 0.02 only made things worse, as MSPE increased to 9.6 with about eight correct and 26 incorrect variables, and so we omitted these results from the table.

Our novel shrinkstage method performs as well or better than all other methods we compared it to. Even with only 20 data sets, the 2-tailed paired Wilcoxon signed rank $p < 0.002$ for comparing MSPE for shrinkstage to that of any of its competitors except the 0.15 and 0.20 level lasso. The 0.20 level lasso ran a close second to shrinkstage, and it is the only method not statistically significantly inferior in terms of MSPE ($p = 0.19$). Still, our shrinkstage procedure selected 77% as many correct genes as the 0.20 level lasso, while selecting about six *times* fewer incorrect genes. Furthermore, the shrinkstage coefficients were typically closer to truth, which is reflected in the estimated 18% reduction in median MSPE.

The shrinkstage method can potentially be further improved, e.g., via judicious choices of the tuning parameters. Altering the tuning parameters of the stepwise approach to choose smaller models might reduce its prediction error somewhat and select fewer incorrect genes, but there is no way to get it to select as many correct genes as either shrinkstage or lasso without running into serious overfitting (and increasing MSPE). For example, merely increasing the p-to-enter for stepwise to just 0.0025 already increased its median MSPE to 13. Thus, we believe there is no way to reduce the MSPE of stepwise selection or univariate screening to the level we have already achieved with shrinkstage.

Table 1
Median (minimum, maximum) performance over 20 simulated data sets with $n=125$, 50% censoring, and 12 truly predictive genes among $p=1,000$ multivariate normal candidate predictors

Method	MSPE	Cor_{lp}	#True	#False	#Selected	$\Sigma\lvert\beta_j\rvert$
Oracle	0.6 (0.1, 3.5)	.99 (.98, 1.00)	12 (12, 12)	0 (0, 0)	12 (12, 12)	13.8 (11.4, 17.7)
Shrinkstage	5.5 (1.1, 10.5)	.79 (.53, .96)	8.5 (4, 12)	5.5 (3, 9)	14 (9, 17)	9.3 (5.1, 11.7)
lasso (.20)	6.7 (4.5, 8.7)	.80 (.65, .90)	11 (9, 12)	32 (22, 38)	43 (34, 49)	8.3 (7.7, 9.4)
Univariate (.01)	8.8 (4.3, 11.6)	.62 (.47, .83)	7 (5, 9)	13.5 (9, 29)	21.5 (14, 35)	7.5 (4.9, 14.6)
Univariate (.005)	9.3 (6.6, 12.4)	.61 (.34, .79)	6 (2, 8)	7 (2, 14)	12.5 (9, 20)	4.6 (2.7, 8.2)
lasso (.05)	9.4 (7.6, 10.8)	.79 (.58, .90)	10 (7, 12)	11 (7, 16)	21 (17, 23)	3.5 (3.0, 4.0)
Stepwise (.002)	9.6 (5.7, 13.9)	.65 (.45, .89)	6 (3, 10)	5 (2, 9)	12 (5, 15)	10.7 (3.6, 21.2)
Empty (null)	14.0 (14.0, 14.0)	.00 (.00, .00)	0 (0, 0)	0 (0, 0)	0 (0, 0)	0.0 (0.0, 0.0)

Although not shown in Table 1, we also began investigating the performance of sequentially adding PCA-derived eignengenes, as well as PLS-type predictors. We observed that even if a large number of PCA-derived eigengenes are included, the models essentially underfit severely, to the point where the MSPE is hardly lower than the empty model as all estimated coefficients are extremely close to zero. The reason for this seems to be that since each eigengene is a linear combination of 988 noise genes and 12 true predictors, the maximum likelihood coefficient estimates have to be very close to 0 so as to keep the contributions from noise genes suppressed. And while PCA seems to essentially never overfit, PLS often overfits in step one. Adding even a single PLS-derived predictor resulted in grossly overestimated coefficients, with correspondingly poor MSPE that was typically worse than the null model. Finally, in our experience the L_2-penalized ridge-regression approach implemented in the `coxph()` function in R and S-plus does not converge in very high-dimensional contexts like these; and even if it did, it would estimate nonzero coefficients for every single gene. Thus, we cannot recommend any of these procedures for use in the context of high-dimensional predictors, at least when the true predictors are relatively low-dimensional and not hugely intercorrelated, as in our simulation context here.

4. Analysis of Breast Cancer Survival

We applied our novel method to van de Vijver et al.'s (9) publicly available breast cancer survival data on $n=295$ subjects with expressions measured on 70 genes that van't Veer et al. (8) previously screened from 24,885 candidate genes. Since data on the vast majority of the 24,885 candidate genes was unavailable to us, we were unfortunately unable to shed any light on which genes beyond the 70 provided might be related to survival. However, even when only considering 70 genes, we observed interesting differences in the results of applying various methods to select the predictive genes and estimate their impact on mortality.

Table 2 displays the relative hazard of mortality per unit change in the standard deviation of the expression of each gene. Only those 32 of the 70 genes that were selected by at least one procedure are shown, whereas the estimated relative hazards are all 1.00 for the other 38 genes. The genes are sorted by their univariate approximate large-sample p-values based on likelihood ratio tests, and the top 16 genes had univariate p-values sufficiently small to satisfy even a 0.05 level test with a Bonferroni correction for 24,885 tests, while 63 of the 70 genes would meet the unadjusted nominal 0.05 level test. The `uni.Bonf` column contains the corresponding

Table 2
Relative hazards per SD of expression for several methods applied to van de Vijver et al.'s breast cancer data on $n = 295$ subjects and 70 genes (screened from 24,885), sorted by univariate p-value. 38 of the 70 genes selected by none of the methods (all relative hazards = 1.00) not shown

Gene	uni.Bonf	ss	ss.adj	lasso	lasso.adj	step	step.adj	step.back
NM.003981LRa	2.10	1.28	1.00	1.05	1.00	1.00	1.00	1.00
NM.016359LRa	2.13	1.76	2.13	1.37	1.40	2.40	2.25	1.00
Contig38288.RCLRa	1.87	1.25	1.00	1.16	1.00	1.00	1.00	1.00
NM.001809LRa	1.98	1.00	1.00	1.00	1.00	1.00	1.00	1.00
Contig55725.RCLRa	1.83	1.00	1.00	1.00	1.00	1.00	1.00	1.00
NM.014321LRa	1.87	1.00	1.00	1.00	1.00	1.00	1.00	1.00
NM.020974LRa	0.51	1.00	1.00	0.95	1.00	1.00	1.00	1.00
NM.004702LRa	1.81	1.00	1.00	1.00	1.00	1.00	1.00	1.00
NM.014791LRa	1.76	1.00	1.00	1.00	1.00	1.00	1.00	1.00
AL137718LRa	1.79	1.00	1.00	1.00	1.00	1.00	1.00	1.00
Contig48328.RCLRa	0.51	1.00	1.00	0.87	1.00	1.00	1.00	1.00
NM.016448LRa	1.83	1.00	1.00	1.04	1.08	1.00	1.00	1.95
Contig28552.RCLRa	1.76	1.00	1.00	1.00	1.00	1.00	1.00	1.00
Contig46218.RCLRa	1.72	1.00	1.00	1.00	1.00	1.00	1.00	1.00
NM.005915LRa	1.70	1.00	1.00	1.00	1.00	1.00	1.00	1.00
NM.000849LRa	0.57	1.00	1.00	1.00	1.00	1.00	1.00	1.00
NM.002916LRa	1.00	1.00	1.00	1.00	1.00	1.00	1.00	1.54
Contig46223.RCLRa	1.00	1.00	1.00	0.96	1.00	1.00	1.00	0.71
NM.020188LRa	1.00	1.00	1.00	1.00	1.00	1.00	1.00	2.03
AF052162LRa	1.00	1.46	1.41	1.16	1.12	1.55	1.47	1.61
X05610LRa	1.00	1.42	1.34	1.19	1.08	1.60	1.44	1.72
Contig63649.RCLRa	1.00	1.34	1.33	1.19	1.13	1.47	1.42	1.39
NM.003875LRa	1.00	1.00	1.00	1.00	1.00	1.00	1.00	0.63
AF201951LRa	1.00	1.00	1.00	0.90	0.95	1.00	1.00	1.00
NM.003862LRa	1.00	1.00	1.00	1.00	1.00	1.00	1.00	1.37
AF257175LRa	1.00	1.00	1.00	1.00	1.00	1.00	1.00	0.59
NM.006117LRa	1.00	1.00	1.00	1.00	1.00	1.00	1.00	1.78
NM.002073LRa	1.00	1.00	1.00	1.00	1.02	1.00	1.00	1.00
AF055033LRa	1.00	1.29	1.38	1.09	1.21	1.37	1.00	1.67
NM.000599LRa	1.00	1.00	1.00	1.00	1.00	1.00	1.42	1.00
Contig32125.RCLRa	1.00	1.00	1.00	1.00	1.00	1.00	1.00	1.36
Contig32185.RCLRa	1.00	1.00	1.00	1.00	1.00	1.00	1.00	0.70

unadjusted relative hazards for each of the top 16 univariately selected genes. Although Bonferroni-corrected tests are often criticized for being overly conservative, we remark that even this might not necessarily be true, in general. The Bonferroni correction is indeed conservative when applied to *exact p*-values based on nonparametric tests, or when each and every *p*-value is exactly uniformly distributed under the null hypothesis, but these conditions are unfortunately seldom met with gene expression data in practice. In the present context, e.g., one would need to trust the accuracy of the large sample approximation based on only 79 events among 295 subjects out to six or more decimal places of the approximate *p*-value in order to begin to have confidence that the Bonferroni correction is actually conservative here, even setting aside the very serious issue of confounding that plagues all univariate methods.

With the exception of `uni.Bonf`, the rest of the estimated relative hazards in Table 2 are all *adjusted* in some fashion for the other selected genes with nonunity relative hazards in the same column, similar to any multivariate Cox proportional hazards regression model. Before considering any genes, we first selected a clinical-only Cox model based only on the four available clinical indicators: ESR1 (1 = estrogen receptor alpha expression, 0 = no), St. Gallen criterion (1 = low risk, 0 = high risk), NIH criterion (1 = low risk, 0 = high risk), and lymph node status from the pathology report (1 = positive, 0 = negative). Both ESR1 (HR = 0.332, $p < 0.000005$) and the St. Gallen (HR = 0.205, $p = 0.036$) indicators were associated with reduced mortality when included in the same model, despite their heavy interdependence given that ESR1 is necessary for St. Gallen, while neither the NIH criterion nor the lymph node status appeared to predict mortality risk, with or without adjusting for other predictors ($p > 0.15$). The `ss.adj` and `ss` columns contain the results of applying our novel shrinkstage algorithm, with and without forcing the ESR1 and St. Gallen clinical indicators into the model, respectively. The five genes selected by both of these procedures were `NM.016359LRa`, `AF052162LRa`, `X05610LRa`, `Contig63649.RCLRa`, and `AF055033LRa`, and we suggest that these particular five genes are thus deserving of further investigation. When the clinical predictors were excluded from the model, `NM.003981LRa` and `Contig38288.RCLRa` were additionally selected by our procedure; our interpretation of this is that the information in these two genes is highly correlated with the two clinical predictors, and thus these genes add very little prognostic information to the available clinical predictors. Of the five genes we selected from the 70 available, adjusted for the clinical predictors, it is interesting but not entirely surprising to us that only one fell in the set of the top 16 univariately ranked genes. Looking at the correlations between `NM.016359LRa` and the other top 15 univariately ranked genes, we found that ten of the 15 pairwise correlations fell between 0.6 and 0.8, indicating a lot of redundant information—not obviously helpful for predicting mortality risk. Finally, we reassessed the independent contributions of ESR1 and the St. Gallen indicator via likelihood ratio-type tests to remove each, while leaving the other in the model, now adjusted for our five selected genes via an offset to force the gene coefficients to remain at their shrinkstage-estimated levels (resulting in a somewhat lower likelihood from the submodels than would be obtained if we allowed the gene coefficients to be reestimated in each submodel). While ESR1 was still significant ($p=0.0077$), the St. Gallen indicator was not ($p=0.45$) unless tested jointly with ESR1 ($p=0.0165$ on 2 df), adjusted for the five selected genes. This implies that whatever predictive information was contained in the St. Gallen indicator was apparently captured by our low-dimensional gene profile along with ESR1. However, since our development of the gene profile was predicated on adjusting for both ESR1 and the St. Gallen criterion, we did not remove St. Gallen but rather simply note the interesting loss of significance, after genetic adjustment. We also reassessed the potential contributions of lmyph node status and the NIH criterion, since it is possible that they could become useful in tandem with our selected gene profile, but there was still no evidence that either improved the model in any way ($p>0.5$).

We programmed a custom stagewise approximation to the lasso procedure in the Cox model framework, with an option to force variables into the model without constraints on their coefficients, and with a default novel stopping rule based on a nominal 0.05 level likelihood ratio test to enter a variable or increase the coefficient on a previously entered variable. We ran this procedure first just on the 70 genes (`lasso`) and again with the two clinical indicators additionally forced into the model (`lasso.adj`). Without adjusting for the clinical predictors, the lasso selected 12 genes, six of which were among the top 16 univariately selected genes, which was rather different from our shrinkstage selections. In fact, the unadjusted lasso selected several genes that no other multivariate method did, with most of them strongly univariately associated with survival. We believe that this is because at the early stages of the procedure the selections are not sufficiently adjusted for other important multivariate predictors, causing the lasso to start off acting much like a univariate selection procedure. Adjusted for the ESR1 and St. Gallen indicators, however, the lasso selected eight genes, with only two among the top 16 univariate genes, and the overall profile was quite similar to `ss.adj` but with a few more selected genes and with more attenuated relative hazard estimates. This fits with our anecdotal simulation experience with the lasso in the Cox model (not shown) in that it appears to over-shrink parameter estimates of true predictors unless the stopping rule is very weak, while selecting too many false predictors unless the stopping rule is very strong. But it appears that adjusting for the relevant clinical predictors did improve the lasso's performance.

We also applied three versions of stepwise Cox regression, all using a nominal 0.01 level likelihood ratio test to stop: `step` represents a stepwise procedure starting from the null model, and without forcing the clinical predictors in the model; `step.adj` starts from the model with the ESR1 and St. Gallen criteria forced into the model; and `step.back` starts from the full model with all 70 genes, with both clinical predictors forced into the model. Interestingly, the unadjusted stepwise search selected exactly the same five genes as `ss.adj`, though four of the five relative hazard estimates were slighlty stronger. Adjusting for the clinical indicators resulted in slightly more attenuated relative hazard estimates for the genes, and `NM.000599LRa` was selected instead of `AF055033LRa`, which hardly matters given that the correlation between these two gene expressions was 0.975 (indicating that either is probably nearly as good a predictor). Starting from the full 72-predictor model and stepping back resulted in selecting 14 genes, eight of which were selected by none of the other procedures. Moreover, this was the *only* procedure not to select `NM.016359LRa`. Overall, we do not believe that `step.back` performed well here, and we cannot recommend it as a general approach; certainly such an approach is never even an option when the number of candidate predictors exceeds the number of uncensored observations. The apparently reasonable performance of the forward stepwise searches is probably in part due to the relatively limited number of available predictors here, as it comes as somewhat of a surprise in comparison with our simulation results that suggest that forward stepwise searches typically miss many important predictors and often select far too many irrelevant genes. In fact, the high concordance of our novel method with the forward stepwise search here leads us to believe that there might be several important predictors among the 24,815 genes that were prescreened and unavailable to us for this analysis.

Finally, our definition of a *prognostic signature* is not merely a set of genes but rather a complete equation that precisely defines a single continuous risk score, based here on the linear predictor of the Cox model (for the log relative hazard). For the `ss.adj` procedure that we advocate, the continuous risk score is calculated as:

`Risk-score`=-0.626(`ESR1`)-0.682(`St.Gallen`)+2.908(`NM.016359LRa`)+1.160(`AF055033LRa`)+2.053(`AF052162LRa`)+1.186(`Contig63649.RCLRa`)+1.844(`X05610LRa`)

Note that the coefficients on the gene expressions in the risk score equation above correspond with the original gene expression scales, not standardized by their sample standard deviations, whereas the relative hazards reported in Table 2 are per unit standard deviation of change in order to better compare the relative impact of each gene. The above risk score could potentially be used to risk stratify future subjects based on these five genes and two clinical indicators, with lower risk scores having superior prognosis—but until independently validated in a future study, caution should be exercised in interpretation. For the 295 subjects used to develop this risk score, the distribution of the risk score appeared approximately normal with mean −0.881, standard deviation 1.264, 75th percentile 0.039, and 95th percentile 1.131. So, roughly speaking, positive risk scores correspond to those in the upper quartile of risk. Van de Vijver et al. ultimately classified each subject as having "good" or "poor" prognosis by thresholding a subject-specific correlation coefficient of their 70-gene profile with the average gene profile of "good outcome patients," defined as all patients with no observed distant metastasis during follow-up, regardless of the length of follow-up. We calculated the sample correlation coefficient between our risk scores and their continuous measure to be −0.783 ($p<0.000001$), and a scatter plot revealed the relationship to be very linear. Thus, although the agreement in prognosis is not perfect and it is unclear which might be superior, the concordance is relatively high, even though our signature is based on 14 times fewer genes. Since the coefficients on all five of our selected genes are positive, higher expressions of these genes appear to correspond with increased risk of mortality. Since the coefficients on both clinical predictors are negative, the ESR1 and the St. Gallen indicators indeed correspond with lower risk, as their definitions imply. Having ESR1 corresponds with a 46.5% reduction in risk, all other gene expressions being the same, while meeting the St. Gallen criteria corresponds with a full 73.0% risk reduction, since any woman satisfying the St. Gallen criteria also has ESR1 (and so the coefficients on `ESR1` and `St.Gallen` in the risk score equation are summed, prior to exponentiating to obtain the relative hazard). Thus, any prognostic signature that ignores these available clinical predictors would seem to be quite incomplete and potentially misleading. However, if our risk score equation reasonably approximates the true log relative hazard, it does imply that up-regulation of three or more of the five risk genes by even one standard deviation could more than offset the protective effects associated with the St. Gallen criterion.

5. Discussion

We introduced a novel method of constructing a prognostic gene signature based on censored survival data, with multivariate adjustments for clinical covariates as well as for multiple selected genes. We chose to reduce the problem to that of model selection and parameter estimation in the Cox proportional hazards model framework in the context of high-dimensional data, where most standard model selection strategies and estimation methods typically break down rather badly. Given this choice of model framework, our procedure inherits both strengths and weaknesses typically associated with the Cox model. For example, if interactions are to be modeled, they must first be explicitly defined by the user and included as additional candidate predictors, and doing this can quickly expand the number of candidate predictors to enormous numbers, making the model selection problem that much harder. In this paper, we chose to assume no interactions, though such enrichments could easily be incorporated.

We compared our procedure to the stagewise approximate lasso, stepwise selection, and to fitting a Cox model to a univariately screened set of predictors. In a simulation setting where the Cox model holds, provided all of the true predictors are included in the model, we demonstrated the clear superiority of our procedure to these competing approaches, all of which were observed to perform relatively poorly in one or more ways. In future work, we plan to investigate wider sets of simulation models and consider stopping rules in greater depth, and we look forward to further improving our method, e.g., given its current dependence on several *ad hoc* tuning parameters.

We applied our novel method to breast cancer survival data, using a data set with a relatively large number of subjects, and we identified a much simpler prognostic signature compared with previous attempts by others. Our profile focuses on just five key genes, and it includes already established clinical predictors, which we argue is the most sensible way to proceed. Our results suggest that profiles based on 70 or more genes might be overly complicated, while only a handful might be needed for effective prognosis. Unfortunately we did not have access to the expression data on the vast majority of the nearly 25,000 genes, and so we are left to wonder how many of them might also be associated with survival, adjusted for clinical predictors and other genes.

Acknowledgments

This work was supported in part by Concept Award W81XWH-04-1-0714 from the Breast Cancer Research Program of the Congressionally Directed Medical Research Programs run by the US Department of Defense via the US Army Medical Research and Materiel Command; the University of Rochester CTSI grant # 1 UL1 RR024160-01 from the National Center for Research Resources (NCRR), a component of the National Institutes of Health (NIH); and the NIH Roadmap for Medical Research.

References

1. Ben-Dor A, Bruhn L, Friedman N, Nachman I, Schummer M, Yakhini Z (2000) Tissue classification with gene expression profiles. J Comput Biol 7:559–584
2. Golub TR, Slonim DK, Tamayo P et al (1999) Molecular classification of cancer: class discovery and class prediction by gene expression monitoring. Science 286:531–537
3. Nguyen DV, Roche DM (2002) Tumor classification by partial least squares using microarray gene expression data. Bioinformatics 18:39–50
4. Park PJ, Tian L, Kohane IS (2002) Linking expression data with patient survival times using partial least squares. Bioinformatics 18:1625–1632
5. Pomeroy SL, Tamayo P, Gaasenbeek M et al (2001) Prediction of central nervous system embryonal tumour outcome based on gene expression. Nature 415:24
6. Rosenwald A, Wright G, Wiestner A et al (2003) The proliferation gene expression signature is a quantitative integrator of oncogenic events that predicts survival in mantle cell lymphoma. Cancer Cell 3:185–197
7. Sorlie T, Perou CM, Tibshirani R et al (2001) Gene expression patterns of breast carcinomas distinguish tumor subclasses with clinical implications. Proc Natl Acad Sci USA 98: 10869–10874
8. van't Veer LJ, Dai H, van de Vijver MJ et al (2002) Gene expression profiling predicts clinical outcome of breast cancer. Nature 415:530–536
9. van de Vijver MJ, He YD, van't Veer LJ et al (2002) A gene-expression signature as a predictor of survival in breast cancer. N Engl J Med 347:1999–2009
10. West M, Blanchette C, Dressman H, Huang E, Ishida S, Spang R et al (2001) Predicting the clinical status of human breast cancer by using gene expression profiles. Proc Natl Acad Sci USA 98:11462–11467
11. Wold H (1966) Estimation of principal components and related models by iterative least squares. In: Krishnaiaah PR (ed) Multivariate analysis. Academic Press, New York, pp 391–420
12. Garthwaite PH (1994) An interpretation of partial least squares. J Am Stat Assoc 89:122–127
13. Hastie T, Tibshirani R, Friedman J (2001) The elements of statistical learning: data mining, inference, and prediction. Springer, New York
14. Li H, Gui J (2004) Partial Cox regression analysis for high-dimensional microarray gene expression data. Bioinformatics 20:1208–1215. doi: 10.1093/bioinformatics/6th900
15. Tibshirani R (1996) Regression shrinkage and selection via the lasso. J R Stat Soc Ser B 58:267–288

Chapter 7

Clustering of Gene Expression Data Via Normal Mixture Models

G.J. McLachlan, L.K. Flack, S.K. Ng, and K. Wang

Abstract

There are two distinct but related clustering problems with microarray data. One problem concerns the clustering of the tissue samples (gene signatures) on the basis of the genes; the other concerns the clustering of the genes on the basis of the tissues (gene profiles). The clusters of tissues so obtained in the first problem can play a useful role in the discovery and understanding of new subclasses of diseases. The clusters of genes obtained in the second problem can be used to search for genetic pathways or groups of genes that might be regulated together. Also, in the first problem, we may wish first to summarize the information in the very large number of genes by clustering them into groups (of hyperspherical shape), which can be represented by some metagenes, such as the group sample means. We can then carry out the clustering of the tissues in terms of these metagenes. We focus here on mixtures of normals to provide a model-based clustering of tissue samples (gene signatures) and of gene profiles.

Key words: Clustering of tissue samples, Clustering of gene profiles, Model-based methods, Normal mixture models, Mixtures of common factor analyzers, Mixtures of linear mixed-effects models

1. Introduction

DNA microarray technology, first described in the mid-1990s, is a method to perform experiments on thousands of gene fragments in parallel. Its widespread use has led to a huge growth in the amount of expression data available. A variety of multivariate analysis methods has been used to explore these data for relationships among the genes and the tissue samples. Cluster analysis has been one of the most frequently used methods for these purposes. It has demonstrated its utility in the elucidation of unknown gene function, the validation of gene discoveries, and the interpretation of biological processes; see refs. (1, 2) for examples.

Andrei Y. Yakovlev et al. (eds.), *Statistical Methods for Microarray Data Analysis: Methods and Protocols*, Methods in Molecular Biology, vol. 972, DOI 10.1007/978-1-60327-337-4_7, © Springer Science+Business Media New York 2013

The main goal of microarray analysis of many diseases, in particular of unclassified cancer, is to identify as yet unclassified cancer subtypes for subsequent validation and prediction, and ultimately to develop individualized prognosis and therapy. Limiting factors include the difficulties of tissue acquisition and the expense of microarray experiments. Thus, often microarray studies attempt to perform a cluster analysis of a small number of tumor samples on the basis of a large number of genes, and can result in gene-to-sample ratios of approximately 100-fold.

Also, many researchers have explored the use of clustering techniques to organize genes into clusters with similar behavior across relevant tissue samples (or cell lines). Although a cluster does not automatically correspond to a pathway, it is a reasonable approximation that genes in the same cluster have something to do with each other or are directly involved in the same pathway. The biological rationale underlying the clustering of the gene profiles is the fact that often many coexpressed genes are also coregulated, which is supported both by an immense body of empirical observations and by detailed mechanistic explanation (3). However, it has been observed that genes with similar profiles sometimes do not share biological similarity (4–6). Thus clustering does not provide proof of relationships between the genes, but it does provide suggestions that help to direct further research.

It can be seen there are two distinct but related clustering problems with microarray data. One problem concerns the clustering of the tissues on the basis of the genes; the other concerns the clustering of the genes on the basis of the tissues. This duality in cluster analysis is quite common.

One of the difficulties of clustering is that the notion of a cluster is vague. A useful way to think about the different clustering procedures is in terms of the shape of the clusters produced (7). The majority of the existing clustering methods assume that a similarity measure or metric is known a priori; often the Euclidean metric is used. But clearly, it would be more appropriate to use a metric that depends on the shape of the clusters. As pointed out by (8), the difficulty is that the shape of the clusters is not known until the clusters have been found, and the clusters cannot be effectively identified unless the shapes are known.

Before we proceed to consider the clustering of microarray data, we give a brief account of clustering in a general context.

2. Cluster Analysis

Cluster analysis is concerned with grouping a number (n) of entities into a smaller number (g) of groups on the basis of observations measured on some variables associated with each entity. It is an exploratory technique that attempts to find groups of observations that have similar values on a set of variables. We let $\mathbf{y}_j = (y_{1j}, \ldots, y_{pj})^T$

be the observation or feature vector containing the values of p measurements $y_{1j},\ldots,y_{pj}$ made on the jth entity ($j=1, \ldots, n$) to be clustered. These data can be organized as a matrix,

$$\mathbf{Y}_{p\times n} = (y_{vj}); \tag{1}$$

that is, the jth column of $\mathbf{Y}_{p\times n}$ is the obervation vector $\mathbf{y}_j$.

In discriminant analysis (supervised learning), the data are classified with respect to g known classes and the intent is to form a classifier or prediction rule on the basis of these classified data for assigning an unclassified entity to one of the g classes on the basis of its feature vector. In contrast to discriminant analysis, in cluster analysis (unsupervised learning) there is no prior information on the group structure of the data or, in the case where it is known that the population consists of a number of classes, there are no data of known origin with respect to the classes. The clustering problem falls into two main categories which overlap to some extent (9).

1. What is the best way of dividing the entities into a given number of groups, where there is no implication that the resulting groups are in any sense a natural division of the data. This is sometimes called dissection or segmentation.
2. What is the best way to find a natural subdivision of the entities into groups? Here by natural clusters, it is meant that the clusters can be described as continuous regions of the feature space containing a relatively high density of points, separated from other such regions by regions containing a relatively low density of points (10). It is therefore intended that natural clusters possess the two intuitive qualities of internal cohesion and external isolation (11).

Sometimes the distinction between the search for naturally occurring clusters as in (2) and other groupings as in (1) is stressed; see, for example, ref. 12. But often it is not made, particularly as most methods for finding natural clusters are also useful for segmenting the data.

Clustering methods can be categorized broadly as being hierarchical or nonhierarchical. With a method in the former category, every cluster obtained at any stage is a merger or split of clusters obtained at the previous stage. Hierarchical methods can be implemented in a so-called agglomerative manner (bottom-up), starting with $g=n$ clusters or in a divisive manner (top-down), starting with the n entities to be clustered as a single cluster.

One of the most popular nonhierarchical methods of clustering is k-means, where "k" refers to the number of clusters to be imposed on the data. It seeks to find $k=g$ clusters that minimize the sum of the squared Euclidean distances between each observation $\mathbf{y}_j$ and its respective cluster mean; that is, it seeks to minimize the trace of W, tr W, where

$$\mathbf{W} = \sum_{i=1}^{g}\sum_{j=1}^{n} z_{ij}(y_j - \overline{y}_i)(y_j - \overline{y}_i)^T \tag{2}$$

is the pooled within-cluster sums of squares and products matrix, and

$$\bar{\mathbf{y}}_i = \frac{\sum_{j=1}^{n} z_{ij}\mathbf{y}_j}{\sum_{j=1}^{n} z_{ij}} \tag{3}$$

is the sample mean of the ith cluster. Here z_{ij} is a zero–one indicator variable that is one or zero, according as $\mathbf{y}_j$ belongs or does not belong to the ith cluster ($i = 1, \ldots, g$; $j = 1, \ldots, n$).

In the absence of any prior knowledge of the metric, it is reasonable to adopt a clustering procedure that is invariant under affine transformations of the data; that is, invariant under transformations of the data of the form,

$$\mathbf{y} \rightarrow \mathbf{C}\mathbf{y} + \mathbf{a}, \tag{4}$$

where C is a nonsingular matrix. If the clustering of a procedure is invariant under (4), for only diagonal C, then it is invariant under change of measuring units but not rotations. But as commented upon in (13), this form of invariance is more compelling than affine invariance. The clustering produced by minimization of |W| is affine invariant.

3. Normal Mixture Models

In this chapter, we shall focus on a model-based approach to the clustering of microarray data using mixtures of normal distributions, which are commonly used in statistics; see, for example, refs. 14–19.

Scott and Symons (20) were one of the first to adopt a model-based approach to clustering. Assuming that the data were normally distributed within a cluster, they showed that their approach is equivalent to some commonly used clustering criteria with various constraints on the cluster covariance matrices. However, from an estimation point of view, this approach leads to inconsistent estimators of the parameters, see, for example, refs. 21, 22.

This inconsistency can be avoided by working with the mixture likelihood formed under the assumption that the observed data are from a mixture of classes corresponding to the clusters to be imposed on the data, as proposed by (23, 24). Finite mixture models have since been increasingly used to model the distributions of a wide variety of random phenomena and to cluster data sets, see, for example, the recent books by Böhning (25), McLachlan and Peel (19), and Frühwirth-Schnatter (26), and the references therein. Earlier references on mixture models may be found in the previous books by Everitt and Hand (27), Titterington et al. (28), McLachlan and Basford (15), and Lindsay (29).

As noted in (30), "Clustering methods based on mixture models allow estimation and hypothesis testing within the framework of standard statistical theory." Previously, Marriott (9, p. 70) had noted that the mixture likelihood-based approach "is about the only clustering technique that is entirely satisfactory from the mathematical point of view. It assumes a well-defined mathematical model, investigates it by well-established statistical techniques, and provides a test of significance for the results." More recently (31) noted that "in the absence of a well-grounded statistical model, it seems difficult to define what is meant by a 'good' clustering algorithm or the 'right' number of clusters."

With the normal mixture mode-based approach to clustering, each observation vector $\mathbf{y}_j$ is assumed to have a g-component normal mixture density,

$$f(\mathbf{y}_j;\Psi) = \sum_{i=1}^{g} \pi_i \phi(\mathbf{y}_j;\mu_i,\Sigma_i), \tag{5}$$

where $\phi(\mathbf{y};\mu_i,\Sigma_i)$ denotes the p-variate normal density function with mean μ_i and covariance matrix Σ_i, and the π_i denote the mixing proportions, which are nonnegative and sum to one. Here the vector Ψ of unknown parameters consists of the mixing proportions π_i, the elements of the component means μ_i, and the distinct elements of the component–covariance matrix Σ_i, and it can be estimated by its maximum likelihood estimate calculated via the EM algorithm; see refs. 21–24. This approach gives a probabilistic clustering defined in terms of the estimated posterior probabilities of component membership $\tau_i(\mathbf{y}_j;\hat{\Psi})$, where $\tau_i(\mathbf{y}_j;\hat{\Psi})$ denotes the posterior probability that the jth feature vector with observed value $\mathbf{y}_j$ belongs to the ith component of the mixture (i=1, …, g; j=1, …, n). Using Bayes' theorem, it can be expressed as

$$\tau_i(\mathbf{y}_j;\Psi) = \frac{\pi_i \phi(\mathbf{y}_j;\mu_i,\Sigma_i)}{\sum_{h=1}^{g} \pi_h \phi(\mathbf{y}_j;\mu_h,\Sigma_h)}. \tag{6}$$

It can be seen that with this approach, we can have a "soft" clustering, whereby each observation may partly belong to more than one cluster. An outright clustering can be obtained by assigning y_j to the component to which it has the greatest estimated posterior probability of belonging. The number of components g in the normal mixture model (5) has to be specified in advance.

One attractive feature of adopting mixture models with elliptically symmetric components such as the normal or its more robust version in the form of the t density (19) is that the implied clustering is invariant under affine transformations (4). Also, in the case where the components of the mixture correspond to externally defined subpopulations, the unknown parameter vector Ψ can be estimated consistently by a sequence of roots of the likelihood equation. Note that this is not the case if a criterion such as minimizing $|W|$ is used.

The normal mixture model-based approach is to be applied here in a nonhierarchical manner, as there is no reason why the clusters of tissues or genes should be hierarchical in nature. It is true that if there is a clear, unequivocal grouping, with little or no overlap between the groups, any method will reach this grouping. But as pointed out by Marriott (9), "hierarchical methods are not primarily adapted to finding groups." As advocated by Marriott (9, p. 67), "it is better to consider the clustering problem *ab initio*, without imposing any conditions."

Another attractive feature of the use of mixture models for clustering is that the question of the number of clusters can be formulated in terms of a criterion or a test for the smallest number of components in the mixture model compatible with the data. One such criterion is the Bayesian information criterion (BIC) of Schwarz (32), while a test can be carried out on the basis of the likelihood ratio statistic λ.

One potential drawback with the normal mixture model-based approach to clustering is that normality is assumed for the cluster distributions. However, this assumption would appear to be reasonable for the clustering of microarray data after appropriate normalization.

4. Clustering of Tissues

Although biological experiments vary considerably in their design, the data generated by microarray experiments can be viewed as a matrix of expression levels. For M microarray experiments (corresponding to M tissue samples), where we measure the expression levels of N genes in each experiment, the results can be represented by a $N \times M$ matrix. For each tissue, we can consider the expression levels of the N genes, called its *expression signature*. Conversely, for each gene, we can consider its expression levels across the different tissue samples, called its *expression profile*. The M tissue samples might correspond to each of M different patients or, say, to samples from a single patient taken at M different time points. The $N \times M$ matrix is portrayed in Fig. 1, where each sample represents a separate microarray experiment and generates a set of N expression levels, one for each gene.

For the clustering of the tissue samples, the microarray data portrayed in Fig. 1 are in the form of the matrix (1) with $n = M$ and $p = N$, and the observation vector $\mathbf{y}_j$ corresponds to the expression signature for the jth tissue sample. In statistics, it is usual to refer to the entirety of the tissue samples as the sample, whereas the biologists tend to refer to each individual expression signature as a sample, and we follow this practice here.

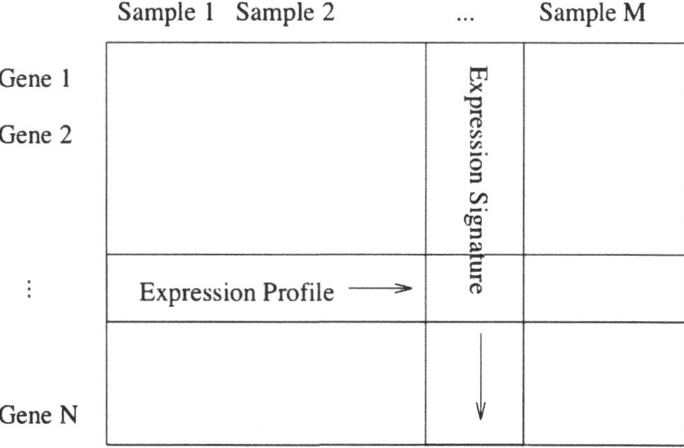

Fig. 1. Gene expression data from M microarray experiments represented as a matrix of expression levels with the N rows corresponding to the N genes and the M columns to the M tissue samples.

4.1. Use of Normal Mixture Models

The normal mixture model (5) cannot be directly fitted to the tissue samples if the number of genes p used in the expression signature is large. This is because the component–covariance matrices Σ_i are highly parameterized with $(1/2)p(p+1)$ distinct elements each. A simple way of proceeding in the clustering of high-dimensional data would be to take the component–covariance matrices Σ_i to be diagonal. But this leads to clusters whose axes are aligned with those of the feature space, whereas in practice the clusters are of arbitrary orientation. For instance, taking the Σ_i to be a common multiple of the identity matrix leads to a soft version of k-means which produces spherical clusters.

Banfield and Raftery (16) introduced a parameterization of the component–covariance matrix Σ_i based on a variant of the standard spectral decomposition of $\Sigma_i (i = 1,\ldots g)$. But if p is large relative to the sample size n, it may not be possible to use this decomposition to infer an appropriate model for the component–covariance matrices. Even if it were possible, the results may not be reliable due to potential problems with near-singular estimates of the component–covariance matrices when p is large relative to n.

Hence, in fitting normal mixture models with unrestricted component–covariance matrices to high-dimensional data, we need to consider first some form of dimension reduction and/or some form of regularization. A common approach to reducing the number of dimensions is to perform a principal component analysis (PCA). However, the latter provides only a global linear model for the representation of the data in a lower-dimensional subspace. Thus it has limited scope in revealing group structure in a data set. A global nonlinear approach can be obtained by postulating a finite mixture of linear (factor) submodels for the distribution of the full observation vector $\mathbf{y}_j$ given a relatively small number of (unobservable) factors. That is, we can provide a local dimensionality reduction method by a mixture of factor analyzers model, which is given by (5) by imposing on the component–covariance matrix Σ_i, the constraint

$$\Sigma_i = \mathbf{B}_i\mathbf{B}_i^T + \mathbf{D}_i \quad (i = 1,\ldots,g), \tag{7}$$

where $\mathbf{B}_i$ is a $p \times q$ matrix of factor loadings and $\mathbf{D}_i$ is a diagonal matrix ($i = 1, \ldots, g$). We can think of the use of this mixture of factor analyzers model as being purely a method of regularization. But in the present context, it might be possible to make a case for it being a reasonable model for the correlation structure between the genes. This model implies that the latter can be explained by the linear dependence of the genes on a small number of latent (unobservable variables) specific to each component.

In practice, there is often the need to reduce further the number of parameters in the specification of the component–covariance matrices.

Baek and McLachlan (33) have proposed therefore a new modified approach that uses common component-factor loadings, which considerably reduces further the number of parameters. Moreover, it allows the data to be displayed in low-dimensional plots.

4.2. EMMIX-GENE Procedure

The EMMIX-GENE program of (34) has been designed for the clustering of tissue samples via mixtures of factor analyzers. In practice we may wish to work with a subset of the available genes, particularly as the fitting of a mixture of factor analyzers will involve a considerable amount of computation time for an extremely large number of genes. Indeed, the simultaneous use of too many genes in the cluster analysis may serve only to create noise that masks the effect of a smaller number of genes. Also, the intent of the cluster analysis may not be to produce a clustering of the tissues on the basis of all the available genes, but rather to discover and study different clusterings of the tissues corresponding to different subsets of the genes (35, 36). As explained in (37), the tissues (cell lines or biological samples) may cluster according to cell or tissue type (for example, cancerous or healthy) or according to cancer type (for example, breast cancer or melanoma). However, the same samples may cluster differently according to other cellular characteristics, such as progression through the cell cycle, drug metabolism, mutation, growth rate, or interferon response, all of which have a genetic basis.

Therefore, the EMMIX-GENE procedure has two optional steps before the final step of clustering the tissues. The first step considers the selection of a subset of relevant genes from the available set of genes by screening the genes on an individual basis to eliminate those which are of little use in clustering the tissue samples. The usefulness of a given gene to the clustering process can be assessed formally by a test of the null hypothesis that it has a single component normal distribution over the tissue samples. Even after this step has been completed, there may still be too many genes remaining. Thus there is a second step in EMMIX-GENE in which the retained gene profiles are clustered (after standardization) into a number of groups on the basis of Euclidean distance so that genes with similar profiles are put into the same group. In general, care has to be taken with the scaling of variables before clustering of the observations, as the nature of the variables can be intrinsically different. In the present context the variables (gene expressions) are measured on the same scale. Also, as noted above, the clustering of the observations (tissues) via normal mixture models is invariant under changes in scale and location. The clustering of the tissue samples can be carried out on the basis of the groups considered individually using some or all of the genes within a group or collectively. For the latter, we can replace each group by a representative (a metagene) such as the sample mean as in the EMMIX-GENE procedure.

4.3. Example: Clustering of Some Prostate Cancer Tissues

To illustrate this approach we applied the EMMIX-GENE procedure to the prostate cancer data of Singh et al. (38). It consists of $n = 6{,}033$ genes and $p = 102$ columns denoting 50 normal tissues and 52 tumors. After applying select-genes to this set, there were 2,208 genes remaining in the set. The remaining genes were clustered into 50 groups, which were ranked on the basis of $-2 \log \lambda$, where λ is the likelihood ratio statistic for testing $g = 1$ versus $g = 2$ components in the mixture model. The tissues were clustered on the basis of the 50 group mean pseudogenes by using a mixture of factor analyzers with four factors. This resulted in a partition of the tissues in which one cluster contains six tumors (11, 26, 47–50) and 44 normals (1–10, 12–25, 27–46) and the other cluster contains 51 (51–75, 77–102) tumors and one normal (76). This corresponds to an error rate of 7 out of 102 tissues compared to the true classification given in (38).

5. Clustering of Gene Profiles

In order to cluster gene profiles, it might seem possible just to interchange rows and columns in the data matrix (1). But with most applications of cluster analysis in practice it is assumed that

(a) There are no replications on any particular entity specifically identified as such.

(b) All the observations on the entities are independent of one another.

These assumptions should hold for the clustering of the tissue samples, although the tissue samples have been known to be correlated for different tissues due to flawed experimental conditions. However, condition (b) will not hold for the clustering of gene profiles, since not all the genes are independently distributed, and condition (a) will generally not hold either as the gene profiles may be measured over time or on technical replicates. While this correlated structure can be incorporated into the normal mixture model (5) by appropriate specification of the component–covariance matrices Σ_i, it is difficult to fit the model under such specifications. For example, the *M*-step (the maximization step of the EM algorithm) may not exist in closed form. Accordingly, we now consider the EMMIX-WIRE model of Ng et al. (39), who adopt conditionally a mixture of linear mixed models to specify this correlation structure among the tissue samples and to allow for correlations among the genes. It also enables covariate information to be incorporated into the clustering process.

5.1. The EMMIX-WIRE Procedure

For a gene microarray experiment with repeated measurements, we have for the *j*th gene ($j = 1, \ldots, n$), when $n = N$, a feature vector (profile vector) $\mathbf{y}_j = (y_{1j}^T, \ldots, y_{tj}^T)^T$, where t is the number of distinct tissues in the experiment and

$$\mathbf{y}_{lj} = (y_{l1j}, \ldots, y_{lrj})^T \quad (l = 1, \ldots, t)$$

contains the r replications on the *j*th gene from the *l*th tissue. Note that here, the r replications can also be time points. The dimension p of the profile vector y_j is equal to the number of microarray experiments, $M = rt$. Conditional on its membership of the *i*th component of the mixture, the EMMIX-WIRE procedure assumes that $\mathbf{y}_j$ follows a linear mixed-effects model (LMM),

$$\mathbf{y}_j = \mathbf{X}\beta_i + \mathbf{U}\mathbf{b}_{ij} + \mathbf{V}\mathbf{c}_i + \varepsilon_{ij}, \tag{8}$$

where the elements of β_i (a t-dimensional vector) are fixed effects (unknown constants) ($i = 1, \ldots, g$). In (8), b_{ij} (a q_b-dimensional vector) and c_i (a q_c-dimensional vector) represent the unobservable gene- and tissue-specific random effects, respectively, conditional on membership of the *i*th cluster. These random effects represent the variation due to the heterogeneity of genes and tissues (corresponding to $\mathbf{b}_i = (b_{i1}^T, \ldots, b_{in}^T)^T$ and c_i, respectively). The random effects b_{ij} and c_i, and the measurement error vector ε_{ij} are assumed to be mutually independent. In (8), X, U, and V are known design matrices of the corresponding fixed or random effects. The dimensions q_b and q_c of the random effects terms b_{ij} and c_i are determined by the design matrices U and V which, along with X and H, specify the experimental design to be adopted.

With the LMM, the distributions of b_{ij} and c_i are taken, respectively, to be multivariate normal $N_{q_b}(0, q_b\mathbf{I}_{q_b})$ and $N_{q_c}(0, q_c\mathbf{I}_{q_c})$, where $\mathbf{I}_{q_b}$ and $\mathbf{I}_{q_c}$ are identity matrices with dimensions being specified by the subscripts. The measurement error vector ε_{ij} is also taken to be multivariate normal $N_p(0, \mathbf{A}_i)$, where $\mathbf{A}_i = \text{diag}(\mathbf{H}\varphi_i)$

is a diagonal matrix constructed from the vector $(\mathbf{H}\varphi_i)$ with $\varphi_i = (\sigma_{i1}^2, \ldots, \sigma_{iq_e}^2)^T$ and H is a known $p \times q_e$ zero–one design matrix. That is, we allow the *i*th component-variance to be different among the *p* microarray experiments.

The vector Ψ of unknown parameters can be obtained by maximum likelihood via the EM algorithm, proceeding conditionally on the tissue-specific random effects $\mathbf{c}_i$. The E- and M-steps can be implemented in closed form. In particular, an approximation to the E-step by carrying out time-consuming Monte Carlo methods is not required. A probabilistic or an outright clustering of the genes into *g* components can be obtained, based on the estimated posterior probabilities of component membership given the profile vectors and the estimated tissue-specific random effects $\hat{c}_i$ $(i = 1, \ldots, g)$.

5.2. Example: Clustering of Some Yeast Profiles

To illustrate this method, we consider the CDC28 dataset, which contains more than 6,000 genes measured at 17 time points (0, 10, 20, …, 160) over 160 min, which is about two periods of yeast cell under CDC28 condition. Cho et al. (40) and Yeung et al. (31) identified and clustered some of the 6,000 genes into different functional groups. For example, Yeung et al. (31) presented 384 genes corresponding to five functional groups, among which there are 237 genes falling into four MIPS functional groups (DNA synthesis and replication, organization of centrosome, nitrogen and sulfur metabolism, and ribosomal proteins). Wong et al. (41) reanalysed the 237 cell cycle data, using their two-stage clustering method and found that it outperformed the other methods that they tried. They were an hierarchical method, *k*-means, SOM, SOTA, and a normal mixture model-based procedure corresponding to (5), which were all used to cluster the 237 genes into $g=4$ clusters. On comparing the latter with the four MIPS functional groups, they reported that the the Rand Index (RI) for their two-stage method was equal to 0.7087. In this paper, we shall compare the EMMIX-WIRE procedure with the two-stage clustering method.

In this example, the gene profile vector y_j for the *j*th gene is given by

$$\mathbf{y}_j = (y_{j1}, \ldots, y_{jp})^T,$$

where y_{jt} denotes the expression level of the *j*th gene at time t $(t=1, \ldots, p)$ and $p=16$. Here $p=16$ and not 17, as the sample at the first time point (that is, immediately after arrest) was eliminated from our analysis.

The model (8) was fitted with $\beta_i = (\beta_{1i}, \beta_{2i})^T$ as the fixed-effects vector for the *i*th component and with the *k*th row of the design matrix X, corresponding to the time point t_k, given by

$$\left(\cos\left(\frac{2\pi t_k}{\omega} + \Psi\right) \quad \sin\left(\frac{2\pi t_k}{\omega} + \Psi\right) \right), \tag{9}$$

Table 1
Values of BIC for various levels of the number of components *g*

The number of components					
2	3	4	5	6	7
10,297	10,125	10,117	10,139	10,161	10,126

for $k = 1, \ldots, 16$, where $t_k = 10k$ $(k = 1, \ldots, 16)$. We used the estimates of the cell cycle period ω and initial phase Ψ as obtained in (42); that is, $\omega = 73\text{min}$ and $\Psi = 0.17\pi$. Note that the initial phase parameter is not identifiable from a single experiment. The design matrix U was taken to be $\mathbf{1}_p$ (that is, $q_b = 1$) with $\mathbf{b}_{ij} = b_{ij}$, the common random effect for all time points shared by the *j*th gene, and $\mathbf{H}_i = \mathbf{I}_p$. The cluster-specific random effect c_i was specified as $\mathbf{c}_i = (c_{i1}, \ldots, c_{ip})^T$ with $q_c = p$ and $V = I_p$. With respect to the error terms, we took $W = I_p$ with $q_e = p$.

Concerning the number of components, we report in Table 1 the values of BIC obtained for various levels of the number of components g. As we were unable to calculate the likelihood exactly under the model (8) in the case of nonzero cluster-specific random-effects terms c_i, we approximated it by taking the gene-profile vectors to be independently distributed in forming the log likelihood in calculating the value of BIC. According to the tabulated values of BIC in Table 1, we should choose $g = 4$ components, which agrees with the number of MIPS functional groups in these genes.

For $g = 4$, we found that the estimated variance θ_{ci} for the cluster-specific random-effects term was equal to 0.080, 0.001, 0.171, and 0.289, which indicates some level of correlation within at least two of the four clusters. The Rand Index and its adjusted value were equal to 0.7808 and 0.5455, which compare favorably to the corresponding values of 0.7087 and 0.3697, as obtained by Wong et al. (41) for their method. On permuting the cluster labels to minimize the error rate of the clustering with respect to the four MIPS functional groups, we obtained an error rate of 0.278. We also clustered the genes into four clusters by not having cluster-specific random-effects terms c_i in (8), yielding lower values of 0.7097 and 0.4396 for the Rand Index and its adjustment. Hence in this example, the use of cluster-specific random-effects terms

Table 2
Summary of clustering results for $g=4$ clusters

Model	Rand Index	Adjusted Rand Index	Error rate
1	0.7848	0.5627	0.278
2	0.7097	0.4396	0.278
3	0.7121	0.3768	0.4051
Wong	0.7087	0.3697	Not available

leads to a clustering that corresponds more closely to the underlying functional groups than without their use. But the error rate in this case without cluster-specific random-effects terms is the same as with these terms included.

The clustering obtained without the cluster-specific random-effects terms is still superior in terms of the Rand Index and its adjusted value to the two-stage method of Wong et al. (41), which was the best on the basis of these criteria in their comparative analysis.

We also fitted the mixed linear model mixture (8) without the sine–cos regression model (9) for the mean, but with a separate (fixed effects) term at each of the $p=17$ time points; that is, we set $X=I_p$ and took β_i to be a p-dimensional vector of fixed effects. We did not include cluster-specific random-effects terms c_i due to their nonidentifiability in this case. This nonregression model gave worse results for the Rand Index and the error rate than with the regression model (8) using the sine–cos curve to specify the mean at a given time point. The results for this nonregression version are listed under Model 3 in Table 2, where the clustering results have been summarized. In this table, Models 1 and 2 correspond to the use of the regression model (8) with and without cluster-specific random-effects terms.

In Figs. 2 and 3, we give the plots of the gene profiles as clustered into $g=4$ clusters as obtained by fitting the mixture of linear mixed models (8) with and without cluster-specific random-effects terms c_i. In Fig. 4, the plots of the gene profiles over all 17 time points are grouped according to their actual functional grouping.

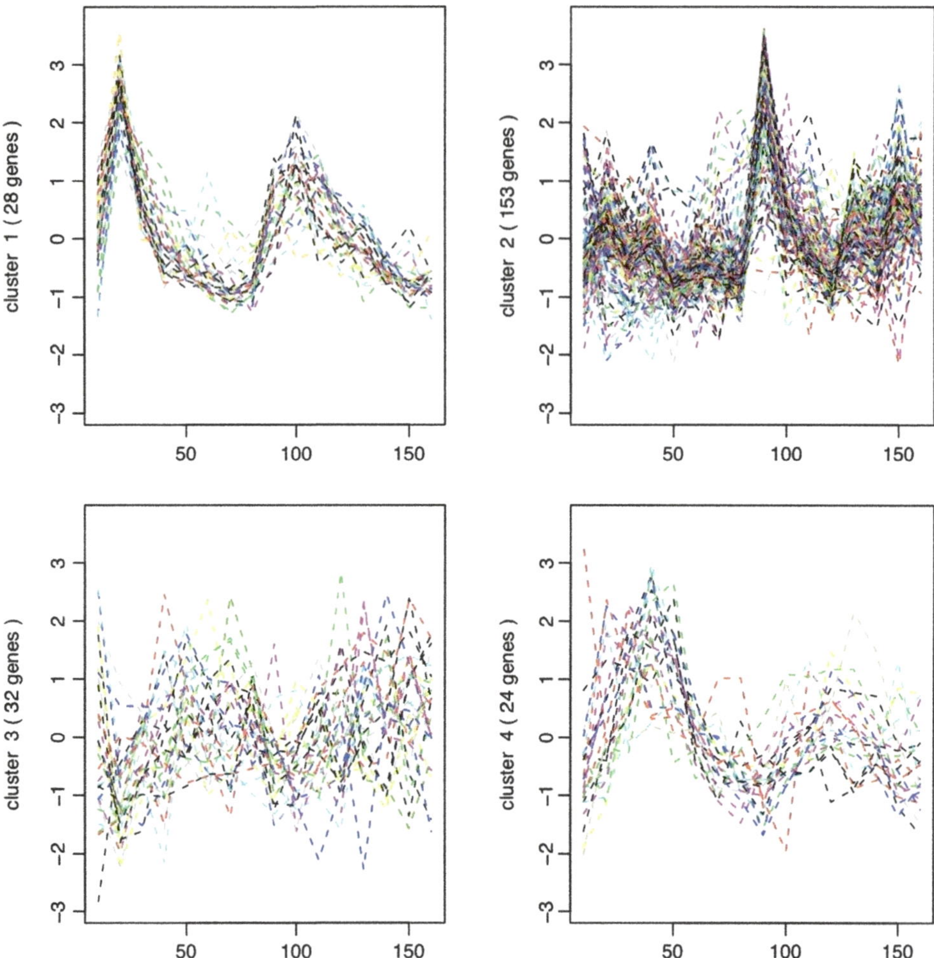

Fig. 2. Clusters of gene profiles obtained by mixture of linear mixed models with cluster-specific random effects.

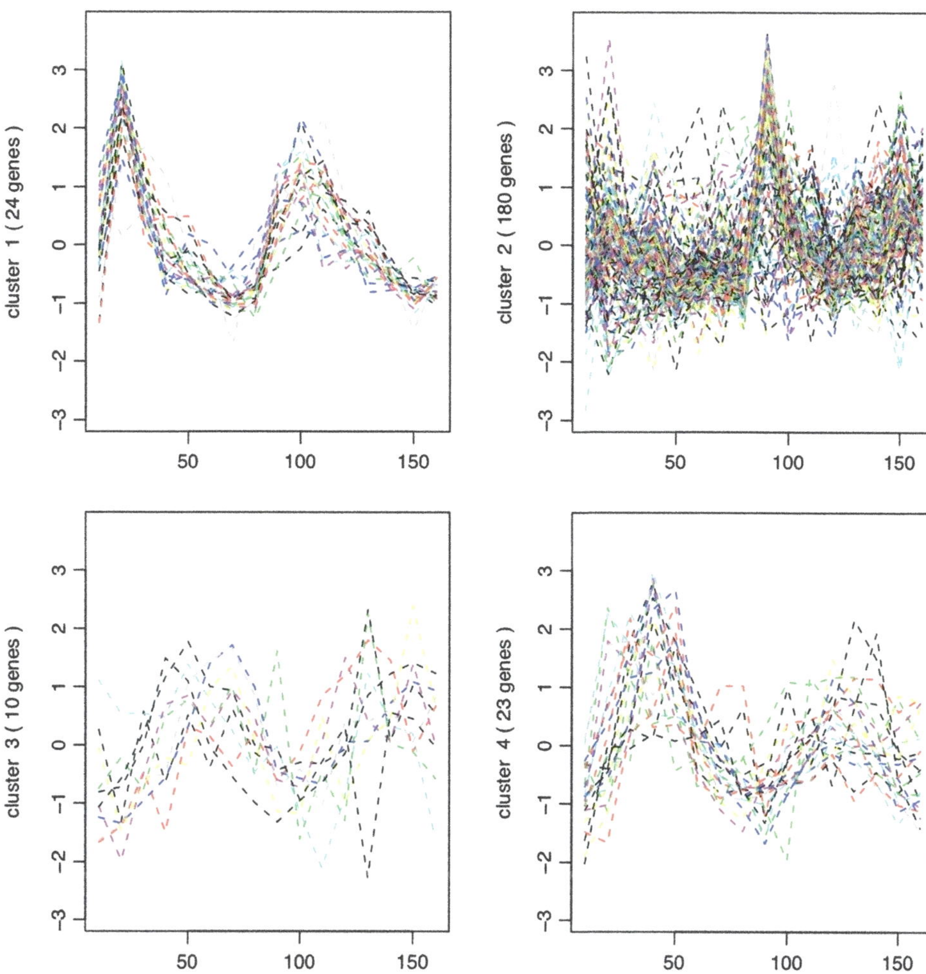

Fig. 3. Clusters of gene profiles obtained by mixture of linear mixed models without cluster-specific random effects.

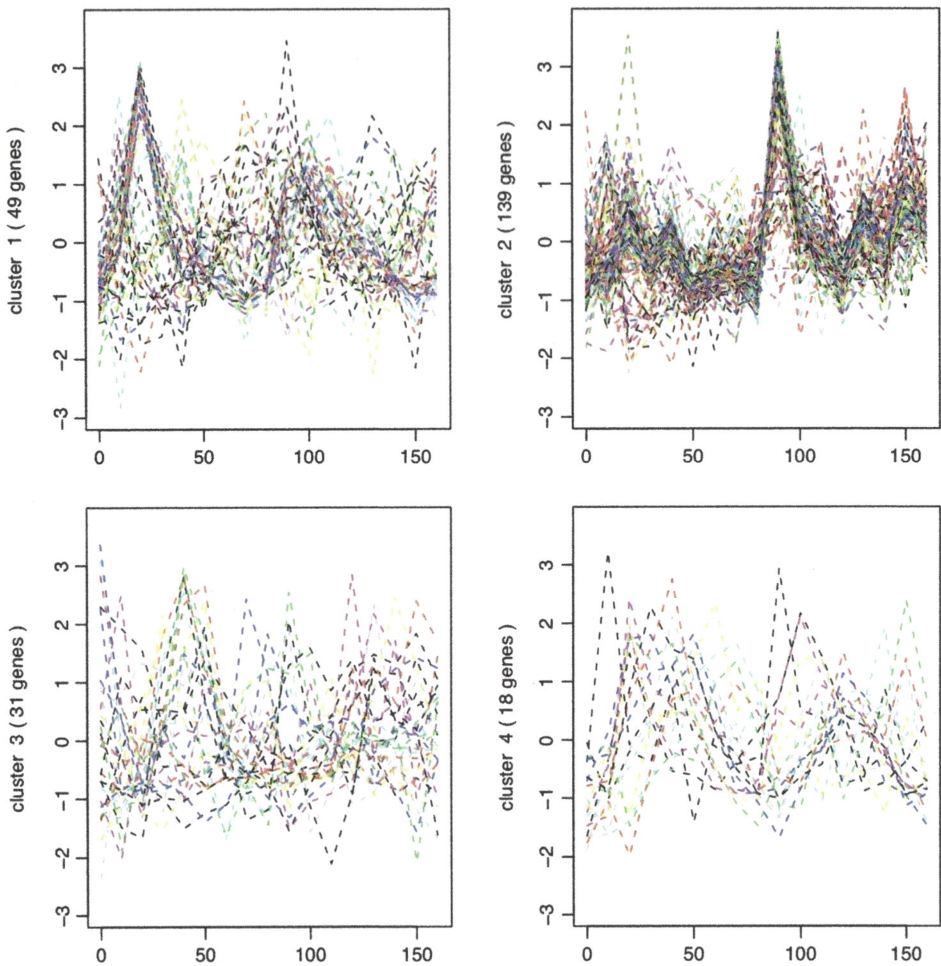

Fig. 4. Plots of gene profiles grouped according to their true functional grouping.

References

1. Alizadeh A, Eisen MB, Davis RE et al (2000) Distinct types of diffuse large B-cell lymphoma identified by gene expression profiling. Nature 403:503–511
2. Eisen MB, Spellman PT, Brown PO, Botstein D (1998) Cluster analysis and display of genome-wide expression patterns. Proc Natl Acad Sci U S A 95:14863–14868
3. Boutros PC, Okey AB (2005) Unsupervised pattern recognition: an introduction to the whys and wherefores of clustering. Brief Bioinform 6:331–343
4. Clare A, King RD (2002) Machine learning of functional class from phenotype data. Bioinformatics 18:160–166
5. Gibbons FD, Roth FP (2002) Judging the quality of gene expression-based clustering methods using gene annotation. Genome Res 12:1574–1581
6. DeRisi JL, Iyer VR, Brown PO (1997) Exploring the metabolic and genetic control of gene expression on a genomic scale. Science 278:680–686
7. Reilly C, Wang C, Rutherford R (2005) A rapid method for the comparison of cluster analyses. Statistica Sinica 15:19–33
8. Coleman D, Dong XP, Hardin J, Rocke DM, Woodruff DL (1999) Some computational issues in cluster analysis with no a priori metric. Comput Stat Data Anal 31:1–11
9. Marriott FHC (1974) The interpretation of multiple observations. Academic, London
10. Everitt BS (1993) Cluster analysis, 3rd edn. Edward Arnold, London

11. Cormack RM (1971) A review of classification (with discussion). J R Stat Soc A 134:321–367
12. Hand DJ, Heard NA (2005) Finding groups in gene expression data. J Biomed Biotechnol 2005:215–225
13. Hartigan JA (1975) Statistical theory in clustering. J Classification 2:63–76
14. Ganesalingham S, McLachlan GJ (1978) The efficiency of a linear discriminant function based on unclassified initial samples. Biometrika 65:658–665
15. McLachlan GJ, Basford KE (1988) Mixture models: inference and applications to clustering. Marcel Dekker, New York, NY
16. Banfield JD, Raftery AE (1993) Model-based Gaussian and non-Gaussian clustering. Biometrics 49:803–821
17. Fraley C, Raferty AE (1998) How many clusters? Which clustering method? Answers via model-based cluster analysis. Comput J 41:578–588
18. Fraley C, Raferty AE (2002) Model-based clustering, discriminant analysis, and density estimation. J Am Stat Assoc 97:611–631
19. McLachlan GJ, Peel D (2000) Finite mixture models. Wiley, New York, NY
20. Scott AJ, Symons MJ (1971) Clustering methods based on likelihood ratio criteria. Biometrics 27:387–397
21. Bryant P, Williamson JA (1978) Asymptotic behaviour of classification maximum likelihood estimates. Biometrika 65:273–281
22. McLachlan GJ (1982) The classification and mixture maximum likelihood approaches to cluster analysis. In: Krishnaiah PI, Kanal I (eds) Handbook of statistics, vol 2. North-Holland, Amsterdam, pp 199–208
23. Wolfe JH (1965) A computer program for the computation of maximum likelihood analysis of types. Research Memo SRM 65-12. U.S. Naval Personnel Research Activity, San Diego
24. Day NE (1969) Estimating the components of a mixture of two normal distributions. Biometrika 56:463–474
25. Böhning D (1999) Computer-assisted analysis of mixtures and applications: meta-analysis, disease mapping and others. Chapman & Hall/CRC, New York, NY
26. Frühwirth-Schnatter S (2006) Finite mixture and Markov switching models. Springer, New York, NY
27. Everitt BS, Hand DJ (1981) Finite mixture distributions. Chapman & Hall, London
28. Titterington DM, Smith AFM, Markov UE (1985) Statistical analysis of finite mixture distributions. Wiley, New York, NY
29. Lindsay BG (1995) Mixture models: theory, geometry and applications. In: NSF-CBMS Regional Conference Series in Probability and Statistics, vol. 5. Institute of Mathematical Statistics and the American Statistical Association, Alexandria, VA
30. Aitkin M, Anderson D, Hinde J (1981) Statistical modelling of data on teaching styles (with discussion). J R Stat Soc A 144:419–461
31. Yeung KY, Fraley C, Murua A, Raftery AE, Ruzzo WL (2001) Model-based clustering and data transformations for gene expression data. Bioinformatics 17:977–987
32. Schwarz G (1978) Estimating the dimension of a model. Ann Stat 6:461–464
33. Baek J, McLachlan GJ (2008). Mixtures of factor analyzers with common factor loadings for the clustering and visualization of high-dimensional data. Technical Report NI08020-HOP, Preprint Series of the Isaac Newton Institute for Mathematical Sciences, Cambridge
34. McLachlan GJ, Bean RW, Peel D (2002) A mixture model-based approach to the clustering of microarray expression data. Bioinformatics 18:413–422
35. Pollard KS, van der Laan MJ (2002) Statistical inference for simultaneous clustering of gene expression data. Math Biosci 176:99–121
36. Friedman JH, Meulman JJ (2004) Clustering objects on subsets of attributes (with discussion). J R Stat Soc B 66:815–849
37. Belitskaya-Levy I (2006) A generalized clustering problem, with application to DNA microarrays. Stat Appl Genet Mol Biol 5, Article 2.
38. Singh D, Febbo P, Ross K, Jackson D, Manola J, Ladd C, Tamayo P, Renshaw A, D'Amico A, Richie J (2002) Gene expression correlates of clinical prostate cancer behaviour. Cancer Cell 1:203–209
39. Ng SK, McLachlan GJ, Wang K, Ben-Tovim JL, Ng S-W (2006) A mixture model with random-effects components for clustering correlated gene-expression profiles. Bioinformatics 22:1745–1752
40. Cho RJ, Huang M, Campbell MJ, Dong H, Steinmetz L, Sapinoso L, Hampton G, Elledge SJ, Davis RW, Lockhart DJ (2001) Transcriptional regulation and function during the human cell cycle. Nat Genet 27:48–54
41. Wong DSV, Wong FK, Wood GR (2007) A multi-stage approach to clustering and imputation of gene expression profiles. Bioinformatics 23:998–1005
42. Booth JG, Casella G, Cooke JEK, Davis JM (2004) Clustering periodically-expressed genes using microarray data: a statistical analysis of the yeast cell cycle data. Technical Report. Department of Biological Statistics and Computational Biology, Cornell University, Ithaca

Chapter 8

Network-Based Analysis of Multivariate Gene Expression Data

Wei Zhi, Jane Minturn, Eric Rappaport, Garrett Brodeur, and Hongzhe Li

Abstract

Multivariate microarray gene expression data are commonly collected to study the genomic responses under ordered conditions such as over increasing/decreasing dose levels or over time during biological processes, where the expression levels of a give gene are expected to be dependent. One important question from such multivariate gene expression experiments is to identify genes that show different expression patterns over treatment dosages or over time; these genes can also point to the pathways that are perturbed during a given biological process. Several empirical Bayes approaches have been developed for identifying the differentially expressed genes in order to account for the parallel structure of the data and to borrow information across all the genes. However, these methods assume that the genes are independent. In this paper, we introduce an alternative empirical Bayes approach for analysis of multivariate gene expression data by assuming a discrete Markov random field (MRF) prior, where the dependency of the differential expression patterns of genes on the networks are modeled by a Markov random field. Simulation studies indicated that the method is quite effective in identifying genes and the modified subnetworks and has higher sensitivity than the commonly used procedures that do not use the pathway information, with similar observed false discovery rates. We applied the proposed methods for analysis of a microarray time course gene expression study of TrkA- and TrkB-transfected neuroblastoma cell lines and identified genes and subnetworks on MAPK, focal adhesion, and prion disease pathways that may explain cell differentiation in TrkA-transfected cell lines.

Key words: Markov random field, Empirical Bayes, KEGG pathways

1. Introduction

Multivariate microarray gene expression data are commonly collected to investigate dose-dependent alterations in gene expression or time-dependent gene expression during a biological process. For example, dose-dependent gene expression data are often measured in the area of toxicology (1, 2) and time-course gene expression data are often

Andrei Y. Yakovlev et al. (eds.), *Statistical Methods for Microarray Data Analysis: Methods and Protocols*, Methods in Molecular Biology, vol. 972, DOI 10.1007/978-1-60327-337-4_8,

collected during a dynamic biological process. For both the dose-dependent and time-course gene expression experiments, the data can be summarized as multivariate vectors, and one goal of such multivariate gene expression studies is to identify genes that have different overall expression patterns between two experiments; these genes can often lead to the identification of the pathways or subnetworks that are perturbed or activated during a given dose-dependent experiment or a dynamic biological process. Compared to gene expression studies of one single experimental condition, such multivariate gene expression data can potentially identify more genes that are differentially expressed (3–6).

One important feature of the multivariate gene expression data is that the data are expected to be dependent across dosages or time points. Efficiently utilizing such dependency can lead to a gain in efficiency in identifying the differentially expressed genes. Yuan and Kendziorski in (3) and Wei and Li (6) developed the hidden Markov model and hidden Markov random field model to identify the differentially expressed genes at each time point for analysis of microarray time-course gene expression data. Instead of identifying genes that are differentially expressed at each time point during a biological process or at a given dosage level, the investigators sometimes are only interested in identifying the genes that show different overall expression patterns during the experiments. Tai and Speed (4) developed an empirical Bayes method treating the observed time-course gene expression data as multivariate vectors. Hong and Li (5) developed a functional empirical Bayes method using B-splines. Both approaches treat the data as multivariate vectors to account for possible correlations of gene expressions over different dosages or time points. These empirical Bayes have approved useful for identifying the relevant genes, they all make the assumptions that the genes are independent with respective to their differential expression states. However, we expect that the differential expression states of genes with transcriptional regulatory relationships are dependent.

The goal of this paper is to model such regulatory dependency by using the prior regulatory network information in order to increase the sensitivities of identifying the biologically relevant pathways. Information about gene regulatory dependence has been accumulated from many years of biomedical experiments and is summarized in the form of pathways and networks and assembled into pathway databases. Some well-known pathway databases include KEGG, BioCarta (www.biocarta.com) and BioCyc (www.biocyc.org). The most common way of utilizing the known regulatory network information in analysis of microarray gene expression data is to first identify the differentially expressed genes using methods, e.g., of (4) or (5) and then to map these genes to the network to visualize which subnetworks show differential expression or to perform some types of gene set enrichment analysis. One limitation of such an approach is that for many multivariate gene expression data sets, the sample sizes are usually small and therefore the approach often has limited power to identify the relevant subnetworks. Representing the known genetic regulatory network as an undirected graph, Wei and Li (7) and Wei and Pan (8) have recently developed hidden Markov random field (MRF)-based models for identifying the subnetworks that show differential expression patterns between two conditions and have demonstrated using both simulations and applications to real data sets that the procedure is more sensitive in identifying the differentially expressed genes than those procedures that do not utilize the pathway structure information. However, neither of these explicitly models the multivariate expression data. Wei and Li in (6) extended the model of Wei and Li (7) and the HMM model of (3) to analyze the microarray time course gene expression in the framework of a hidden spatial–temporal MRF model. However, this approach aims to identify the differentially expressed genes at each time point and it assumes the same network dependency of the gene differential expression states at all the time points.

In this paper, to efficiently identify the differentially expressed genes in the multivariate gene expression experiments, we develop the hidden MRF model of (7) further into a hidden MRF model for multivariate gene expression data in order to take into account potential dependency of gene expression over time and the known biological pathway information. We treat the multivariate gene expression data as multivariate data, allowing for dependency of the data across the dosage levels or over time points. Different from the popular empirical Bayes methods for analysis of multivariate gene expression data where genes and their differentially expression states are assumed to be independent, this method models the dependency of the differentially expression states using a discrete Markov random field and therefore enables the information of a known network of pathways to be efficiently utilized in order to identify more biologically interpretable results. Although the formulation of the problem is similar to that of (7), models for multivariate gene expression data are more complicated and require new methods for estimating the model parameters. We propose to use both the moment estimate and maximum likelihood estimates in the iterative conditional mode (ICM) algorithm (9, 10).

We first introduce the hidden MRF model for multivariate expression data and present an efficient algorithm for parameter estimation by the ICM algorithm. We then present results from simulation studies to demonstrate the application of the hidden MRF model, to compare with existing methods, and to evaluate the sensitivity of the method to misspecification of the network structure. For a case study, we apply the hidden MRF model to analyze the time-course gene expression data of TrkA- and TrkB-transfected neuroblastoma cell lines in order to identify the pathways that are related to cell differentiation in TrkA-transfected cell lines. Finally, we present a brief discussion of the methods.

2. Statistical Models and Methods

We first introduce a hidden MRF model for multivariate gene expression data, where the network structure is represented as an undirected graph. The model is an extension of the model of (7) to multivariate gene expression data, where the distribution of latent differential states of the genes is modeled as a discrete MRF defined on the prior network structure, and the empirical Bayes model of (4) are used for modeling the emission density for the observed multivariate gene expression data.

2.1. Data Observed and Representation of Genetic Networks as Undirected Graphs

Consider the multivariate gene expression data measured under two different conditions over k dosage levels or time points, with n independent samples measured under one condition and m independent samples measured under another condition. For each experiment, we assume that the expression levels of p genes are measured. For a given gene g, we denote these data as *i.i.d.* $k\times 1$ random vectors $\mathbf{Y}_{g1},\ldots,\mathbf{Y}_{gn}$ for condition 1 and $\mathbf{Z}_{g1},\ldots,\mathbf{Z}_{gm}$ for condition 2. We further assume that $\mathbf{Y}_{gi} \sim N_k(\mu_{gy},\Sigma_g)$ and $\mathbf{Z}_{gi} \sim N_k(\mu_{gz},\Sigma_g)$. For a given gene g, the null hypothesis of interest is

$$H_{g0}:\mu_{gy}=\mu_{gz}. \tag{1}$$

Define $\mu_g=\mu_{gy}-\mu_{gz}$. For a given gene g, let I_g take the value of 1 if $\mu_g \neq 0$ and 0 if $\mu_g = 0$. We call the genes with $I_g=1$ the differentially expressed (DE) genes. Our goal is to identify these DE genes among the p genes.

Besides the gene expression data, suppose that we have a network of known pathways that can be represented as an undirected graph $G=(V, E)$, where V is the set of nodes that represent genes or proteins coded by genes and E is the set of edges linking two genes with a regulatory relationship. Let $p=|V|$ be the number of genes that this network contains. Note the gene set V is often a subset of all the genes that are probed on the gene expression arrays. If we want to include all the genes that are probed on the expression arrays, we can expand the network graph G to include isolated nodes, which are those genes that are probed on the arrays but are not part of the known biological network. For two genes g and g', if there is a known regulatory relationship, we write $g\sim g'$. For a given gene g, let $N_g=\{g':g\sim g'\in E\}$ be the set of genes that have a regulatory relationship with gene g and $d_g=|N_g|$ be the degree for gene g.

2.2. A Discrete Markov Random Field Model for Differential Expression States for Genes on the Network

Our goal is to identify the genes on the network G that are multivariate differentially expressed between the two experimental conditions. Since two neighboring genes g and g' have regulatory relationship on the network, we should expect that the DE states I_g and $I_{g'}$ are dependent. In order to model the dependency of I_g

over the network, following Wei and Li (7), we introduce a simple MRF model. Particularly, we assume the following auto-logistic model for the conditional distribution of I_g,

$$\Pr(I_g \mid I_{g'}, g' \neq g) = \frac{\exp\{I_g F(I_g)\}}{1+\exp\{F(I_g)\}}, \tag{2}$$

where

$$F(I_g) = \gamma + \beta \frac{\sum_{g' \in N_g} (2I_{g'} - 1)}{d_g},$$

and γ and $\beta \geq 0$ are arbitrary real numbers. Here the parameter β measures the dependency of the differential expression states of the neighboring genes. We assume that the true DE states $(I^*_g) = \{I^*_g, g = 1,\ldots,p\}$ is a particular realization of this locally dependent MRF. Note that when $\beta = 0$, the model assumes that all the I_gs are independent with the same prior probability $\exp(\gamma)/(1+\exp(\gamma))$ of being a DE gene.

2.3. Emission Probabilities for Multivariate Gene Expression Data and the HMRF Model

To relate the differential expression state I_g to the observed gene expression data $\mathbf{D}_g = (\mathbf{Y}_{g1},\ldots,\mathbf{Y}_{gn};\mathbf{Z}_{g1},\ldots,\mathbf{Z}_{gm})$, we follow the empirical Bayes approach of Tai and Speed (4) for multivariate gene expression data and use conjugate priors for μ_g and Σ_g, that is, an inverse Wishart prior for Σ_g and a dependent multivariate normal prior for μ_g. To make notation simple, we drop the gene subscript g when introducing the Bayesian model. Let

$$\bar{\mathbf{Y}} = (\mathbf{Y}_1 + \cdots + \mathbf{Y}_n)/n, \quad \bar{\mathbf{Z}} = (\mathbf{Z}_1 + \cdots + \mathbf{Z}_m)/m, \quad \bar{\mathbf{X}} = \bar{\mathbf{Y}} - \bar{\mathbf{Z}},$$

$$\mathbf{S}_y = (n-1)^{-1}\sum_{i=1}^{n}(\mathbf{Y}_i - \bar{\mathbf{Y}})(\mathbf{Y}_i - \bar{\mathbf{Y}})',$$

$$\mathbf{S}_z = (m-1)^{-1}\sum_{i=1}^{m}(\mathbf{Z}_i - \bar{\mathbf{Z}})(\mathbf{Z}_i - \bar{\mathbf{Z}})', \quad \mathbf{S} = (n+m-2)^{-1}((n-1)\mathbf{S}_y + (m-1)\mathbf{S}_z).$$

Following (4), we assign independent and identical inverse Wishart priors to Σ, $\Sigma \sim W^{-1}((\nu\Lambda)^{-1}, \nu)$. Given Σ, we assign multivariate normal priors for the gene-specific mean difference μ for the two cases $(I=1)$ and $(I=0)$:

$$\mu \mid \Sigma, I = 1 \sim N_k(0, \eta^{-1}\Sigma),$$

$$\mu \mid \Sigma, I = 0 \equiv 0.$$

Since the statistics $(\bar{\mathbf{X}}, \mathbf{S})$ are the sufficient statistics for the parameters (μ, Σ) (Tai and Speed (4)), the conditional distribution of the data $\mathbf{D} = (\mathbf{Y}_1,\ldots,\mathbf{Y}_n,\mathbf{Z}_1,\ldots,\mathbf{Z}_m)$ can be written as

$$P(\mathbf{D} \mid I) = P(\mathbf{Y}_1,\ldots,\mathbf{Y}_n,\mathbf{Z}_1,\ldots,\mathbf{Z}_m \mid I) = P(\bar{\mathbf{X}}, \mathbf{S} \mid I).$$

Tai and Speed further derived in (4)

$$P(\mathbf{D} \mid I = 1) = \frac{\Gamma_k((N+\nu)/2)}{\Gamma_k((N-1)/2)\Gamma_k(\nu/2)} \times (N-1)^{\frac{k(N-1)}{2}} \nu^{-\frac{kN}{2}} (\pi(n^{-1}+m^{-1}+\eta^{-1}))^{-\frac{k}{2}} \times \frac{|\Lambda|^{-\frac{N}{2}} |\mathbf{S}|^{\frac{(N-k-2)}{2}}}{|\mathbf{I}_k + ((n^{-1}+m^{-1}+\eta^{-1})\nu\Lambda)^{-1}\overline{\mathbf{X}}\overline{\mathbf{X}}' + S^*|^{\frac{(N+\nu)}{2}}} \tag{3}$$

where $N = n + m - 1$ and $S^*(\nu\Lambda/(N-1))^{-1}\mathbf{S}$. Thus, given $I = 1$, the probability density function of the data is a function of $\mathbf{X}$ and $\mathbf{S}$ only, which follows a Student-Siegel distribution (see ref. 11). Following (11) and (4)'s notation, this distribution is denoted by $StSi_k(\nu, 0, (n^{-1}+m^{-1}+\eta^{-1})\Lambda, N-1, (N-1)^{-1}\nu\Lambda)$. Similarly, the distribution of $P(\mathbf{D}|I=0)$ follows $StSi_k(\nu, 0, (n^{-1}+m^{-1})\Lambda, N-1, (N-1)^{-1}\nu\Lambda)$ with the following density function

$$P(\mathbf{D} \mid I = 0) = \frac{\Gamma_k((N+\nu)/2)}{\Gamma_k((N-1)/2)\Gamma_k(\nu/2)} \times (N-1)^{\frac{k(N-1)}{2}} \nu^{-\frac{kN}{2}} (\pi(n^{-1}+m^{-1}))^{-\frac{k}{2}} \times \frac{|\Lambda|^{-\frac{N}{2}} |\mathbf{S}|^{\frac{N-k-2}{2}}}{|\mathbf{I}_k + ((n^{-1}+m^{-1})\nu\Lambda)^{-1}\overline{\mathbf{X}}\overline{\mathbf{X}}' + S^*|^{\frac{N+\nu}{2}}}. \tag{4}$$

Together the transition probability (2) and the emission probabilities (3) and (4) define a hidden MRF model for multivariate gene expression data with parameters in the emission probabilities $\Theta = (\eta, \nu, \Lambda)$. Define $(I_g) = \{I_1, \ldots, I_p\}$ to be a vector of the differential expression states of the p genes on the network. By Bayes rule, $\Pr(I_g \mid \mathbf{D}) \propto \Pr(\mathbf{D} \mid I_g) \times \Pr(I_g)$. The estimate $(\hat{I}_g)$ that maximizes $\Pr(I_g \mid \mathbf{D})$ is a maximum a posterior (MAP) estimate under 0–1 loss. In order to estimate the parameters and (I_g), we make the following conditional independence assumption.

Assumption: Given any particular realization (I_g), the random variables $(\mathbf{D}) = (\mathbf{D}_1, \mathbf{D}_2, \ldots, \mathbf{D}_g)$ are conditionally independent and each $\mathbf{D}_g$ has the same unknown conditional density function $P(\mathbf{D}_g \mid I_g)$, dependent only on I_g. The conditional density of the observed gene expression data $\mathbf{D}$, given $\mathbf{G}$ and parameter $\Theta = (\eta, \nu, \Lambda)$, is simply,

$$L_\Theta((\mathbf{D}_g) \mid (I_g)) = \prod_{g=1}^{p} P(\mathbf{D}_g \mid I_g), \tag{5}$$

where $P(\mathbf{D}_g \mid I_g)$ is defined as (3) or (4).

3. Estimation of the Model Parameters and the Posterior Probabilities of the DE States

When inferring $(I_g)^*$, parameter estimation must be carried out simultaneously. We propose the following ICM algorithm of Besag (10) to simultaneously estimate the parameter $\Theta = (\eta, \nu, \Lambda)$ with a positive definite constraint on the covariance matrix Λ in the emission probability model and the parameter $\Phi = (\gamma, \beta)$ in the auto-logistic model. Simultaneously estimating the covariance matrix Λ is difficult due to the fact that its estimate has to be positive definite. We propose to first estimate Λ' using the moment estimator in (4). Specifically, by the weak law of large numbers, $\bar{\mathbf{S}}$ converges in probability to $(\nu - k - 1)^{-1}\nu\Lambda$. Therefore, Λ can be estimated by $\Lambda' = \hat{\nu}^{-1}(\nu - k - 1)\bar{\mathbf{S}}$, where $\hat{\nu} = \max(\text{mean}(\hat{\nu}_j), k + 6), j = 1, \ldots, k$, and $\hat{\nu}_j$ is the estimated prior degrees of freedom based on the jth diagonal elements of the gene-specific sample variance–covariance matrices using the method proposed in Section 6.2 in (12). We then fix Λ at its estimate and estimate the other model parameters within the following ICM algorithm (10), which involves the following iterative steps:

S1. Obtain an initial estimate $(\hat{I}_g)$ of the true state $(I_g)^*$, using simple two sample Hotelling's T_2 test.

S2. Estimate Θ by the value $\hat{\theta}$ which maximizes the likelihood $L_\Theta(\mathrm{D} \mid (\hat{I}_g))$.

S3. Estimate Φ by the value Φ which maximizes the following pseudo-likelihood

$$L_\Phi((\hat{I}_g)) = \prod_{g=1}^{G} \frac{\exp\{I_g F((\hat{I}_g))\}}{1 + \exp\{F((\hat{I}_g))\}}.$$

S4. Carry out a single cycle of ICM based on the current $(\hat{I}_g), \hat{\theta}$ and $\widehat{\Phi}$, to obtain a new $(\hat{I}_g)$: for $g = 1$ to p, update I_g which maximizes

$$P(I_g \mid \mathbf{D}, \hat{I}_{g'}, g' \neq g) \propto P(\mathbf{D}_g \mid I_g; \hat{\theta}) P(I_g \mid \hat{I}_{g'}, g' \neq g; \widehat{\Phi}).$$

S5. Go to step 2 until there is convergence in the estimates.

In step 2, $\Theta = (\eta, \nu)$ in the HMRF model and they can be estimated using any numerical optimization procedure. After the convergence of the ICM algorithm and obtaining the parameter estimates, we can run the Gibbs sampling to obtain the estimate of the posterior probability of $\Pr(I_g = 1 \mid \text{data})$ for each of the gene g. These posterior probabilities can then be used for selecting the DE genes.

4. Simulation Study

We performed simulation studies to evaluate the proposed method and to compare results with other methods for identifying the DE genes. Following (7), we first obtained 33 human regulatory pathways from the KEGG database (December 2006), where we retained only gene–gene regulatory relations. These 33 regulatory pathways are inter-connected and formed a network of regulatory pathways. We represent such a network as an undirected graph where each node is a gene and two nodes are connected by an edge if there is a regulatory relation between corresponding genes. Loops (nodes connected to themselves) were eliminated. This results in a graph with 1,668 nodes and 8,011 edges.

To simulate the differential expression states of the genes on this network, we initialized the genes in the K pathways to be DE and the rest genes to be EE, which gives us the initial G_0. We then performed sampling five times based on the current gene differential expression states, according to the Markov random field model with $\gamma_0 = \gamma_1 = 1$ and $\beta = 2$ (see ref. 7). We chose $K = 5, 9, 13, 17$ to obtain different percentages of genes in DE states. After obtaining the differential expression states for the genes, we simulated the multivariate gene expression levels based on the empirical Bayes models, using the same parameters as in (4): $\eta = 0.5, \nu = 13$ and $\Lambda = A \times 10^{-3}$, where

$$A = \begin{pmatrix} 14.69 & 0.57 & 0.99 & 0.40 & 0.55 & 0.51 & -0.23 \\ 0.57 & 15.36 & 1.22 & 0.84 & 1.19 & 0.91 & 0.86 \\ 0.99 & 1.22 & 14.41 & 2.47 & 1.81 & 1.51 & 1.07 \\ 0.40 & 0.84 & 2.47 & 17.05 & 2.40 & 2.32 & 1.33 \\ 0.55 & 1.19 & 1.81 & 2.40 & 15.63 & 3.31 & 2.75 \\ 0.51 & 0.91 & 1.51 & 2.32 & 3.31 & 13.38 & 3.15 \\ -0.23 & 0.86 & 1.07 & 1.33 & 2.75 & 3.15 & 12.90 \end{pmatrix}.$$

For each condition, we chose the number of independent replications to be 3 for each group and repeated the simulation 100 times.

4.1. Comparison with the Method of Tai and Speed

We first examined the parameter estimates of $\Theta = (\eta, \nu)$ using three different methods: the empirical Bayes (EB) method of Tai and Speed (4), the ICM algorithm incorporating the network structures and the ICM algorithm assuming that all the nodes are singletons (i.e., no dependency of the differential expression states). The performance results are shown in Table 1. We observed that both ICM algorithms provide better estimates of both η and ν than the EB algorithm.

Table 1
Comparison of parameter estimates of three different procedures for four sets of simulations with different percentages of DE genes (*q*)

Method	Parameter	Percentage of DE genes (*q*)			
		0.12 (0.005)	0.19 (0.008)	0.36 (0.009)	0.49 (0.008)
HMRF	$\hat{\eta}$	0.38 (0.026)	0.40 (0.028)	0.41 (0.018)	0.44 (0.020)
	$\hat{\nu}$	13.01 (0.061)	13.06 (90.061)	13.12 (0.059)	13.18 (0.057)
HMRF-I	$\hat{\eta}$	0.31 (0.019)	0.34 (0.017)	0.37 (0.013)	0.39 (0.012)
	$\hat{\nu}$	13.03 (0.060)	13.06 (0.060)	13.15 (0.057)	13.22 (0.056)
EB	$\hat{\eta}$	0.067 (0.004)	0.053 (0.003)	0.042 (0.002)	0.039 (0.001)
	$\hat{\nu}$	7.27 (0.21)	7.43 (0.21)	7.86 (0.23)	8.21 (0.25)

HMRF the proposed HMFR model and the ICM algorithm using the network structures, *HMRF-I* the proposed HMFR model and the ICM algorithm without using the network structures, *EB* the empirical Bayes method of of Tai and Speed (2006). Parameter estimates are averages over 100 simulations; standard error is shown in parentheses. The true parameters are $(\eta, \nu) = (0.5, 13)$

We then compare the sensitivity, specificity, and FDR in identifying the DE genes with the EB method of (4). Since the EB method only provides ranks of the genes and does not infer gene states, for the purpose of comparison, we chose a cutoff value to declare genes to be DE using their method so that their approach would have the closest observed FDR levels to our proposed method. We applied the HMRF model to the simulated data sets. The results are summarized in Table 2, clearly showing that our approach obtained significant improvement in sensitivity compared to the other approaches making an independence assumption of genes. The smaller *p* was, the more improvements we obtained. At the same time, our approach also achieved lower FDRs and comparable specificity. Our proposed algorithm assuming that the genes are independent give very similar results to the EB method of (4).

4.2. Sensitivity to Misspecification of the Network Structure

Due to the fact that our current knowledge of biological networks is not complete, in practice, it is possible that the network structures that we use for network-based analysis are misspecified. The misspecification can be due to either the true edges of the networks being missed or the wrong edges being included in the network, or both of these scenarios. We performed simulation studies to evaluate how sensitive the results of the HMRF approach are to these three types of misspecifications of the network structures. We used the same data sets of 100 replicates as in the previous section but used different misspecified network structures when we fitted the hidden MRF model.

Table 2
Comparison of performance in terms of sensitivity, specificity, and false discovery rate (FDR) of three different procedures based on 100 replications for four different scenarios with different percentages of DE genes (*q*)

q	Method	Sensitivity	Specificity	FDR
0.115 (0.005)	HMRF	0.80 (0.029)	1.00 (0.0023)	0.045 (0.019)
	HMRF-I	0.70 (0.042)	0.99 (0.0027)	0.079 (0.025)
	EB	0.69 (0.054)	0.99 (0.0027)	0.079 (0.05)
0.189 (0.008)	HMRF	0.87 (0.033)	0.99 (0.0049)	0.058 (0.020)
	HMRF-I	0.76 (0.03)	0.99 (0.004)	0.074 (0.018)
	EB	0.75 (0.032)	0.99 (0.0041)	0.075 (0.018)
0.357 (0.009)	HMRF	0.91 (0.016)	0.97 (0.0065)	0.054 (0.010)
	HMRF-I	0.84 (0.020)	0.97 (0.0063)	0.066 (0.011)
	EB	0.83 (0.022)	0.97 (0.0064)	0.066 (0.011)
0.486 (0.008)	HMRF	0.95 (0.012)	0.94 (0.012)	0.061 (0.012)
	HMRF-I	0.88 (0.015)	0.95 (0.0086)	0.060 (0.0093)
	EB	0.88 (0.015)	0.95 (0.0087)	0.060 (0.0094)

HMRF the proposed HMFR model using the network structures, *HMRF-I* the proposed HMFR model without using the network structures, *EB* the empirical Bayes method of Tai and Speed (4) with FDRs matched to the HMRF algorithm. Summaries are averaged over 100 simulations; standard deviation is shown in parentheses

For the first scenario, we randomly removed 801 (10%), 2,403 (30%) and 4,005 (50%), respectively, from the 8,011 true edges from the true KEGG networks when we fit the hidden MRF model. For the second scenario, we randomly added approximately 801, 2,403, and 4,005 new edges to the KEGG network, respectively. Finally, for the third scenario, we randomly selected 90, 70, and 50% of the 8,011 true edges and also randomly added approximately 801, 2,403, and 4,005 new edges to the network, respectively, so that the total number of edges remains approximately 8,011. The results of the simulations over 100 replications are summarized as Fig. 1. First, as expected, since the true number of DE genes is small, the specificities of the HMRF procedure remain very high and are similar when the true network structure is used. Second, we also observed that the FDR rates also remain almost the same as when the true structure is used. However, we observed some decreases in sensitivity in identifying the true DE genes. It is worth pointing out even when the network structure is largely misspecified as in scenario 3, the results from the HMRF model are still comparable to those obtained from the HMRF-I approach where the network structure is not utilized.

Finally, we also applied these simulated data with a randomly created network structure with the same number of nodes and edges. As expected, in this case, the estimate of the β parameter

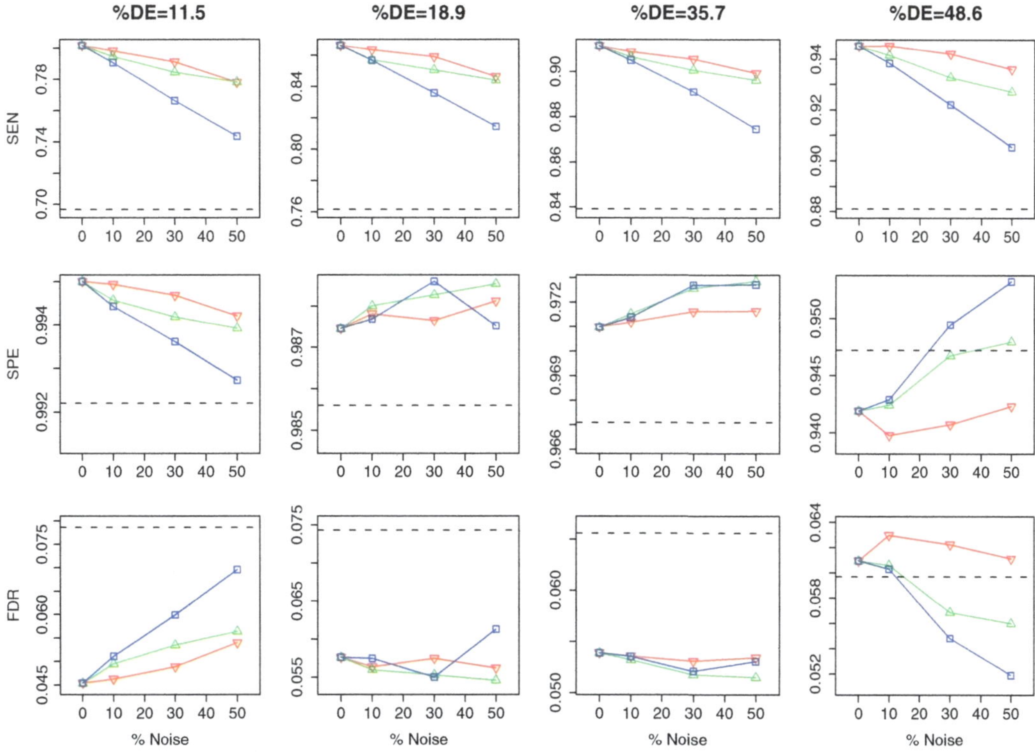

Fig. 1. Results in sensitivity, specificity, and false discovery rate when the network structure is misspecified for four different sets of simulations corresponding to different proportions of DE genes. *Inverted triangle* randomly deleting 10, 30, and 50% of the true edges of the network; *triangle* randomly adding approximately 801 (10%), 2,403 (30%), and 4,005 (50%) new edges to the network; *square* randomly choosing 90, 70, and 50% of the true edges and randomly adding 10, 30, and 50% new edges to the network. The *dashed line* represents results without using the network structures.

was always 0 or very close to 0, and therefore, the results in sensitivity, specificity, and FDR are essentially the same as the method that does not utilize the network structure. These simulations seem to indicate that the results of the HMRF model are not too sensitive to the misspecification of the network structure unless the structure is greatly misspecified.

5. Application to Time-Course Gene Expression Study of TrkA- and TrkB-Transfected Neuroblastoma Cell Lines

Neuroblastoma is the most common and deadly solid tumor in children, but this tumor also has a very high propensity to undergo spontaneous differentiation or regression. Evidence suggests that the Trk family of neurotrophin receptors plays a critical role in tumor behavior (13). Neuroblastomas expressing TrkA are biologically favorable and prone to spontaneous differentiation or regression. In contrast, neuroblastomas expressing TrkB usually have MYCN amplification and are among the most aggressive and deadly tumors known. These tumors also express the TrkB ligand, resulting in an autocrine survival pathway. Unlike the TrkA-expressing tumors, exposure to ligand promotes survival under adverse conditions, but does not cause differentiation. In order to explore the biological basis for the very different behavior of neuroblastomas expressing these highly homologous neurotrophin receptors, a microarray time-course gene expression study was conducted by transfecting TrkA and TrkB into SH-SY5Y cells, a neuronal subclone from the NB cell line SK-N-SH. In particular, full length TrkA and TrkB were cloned into the retroviral expression vector pLNCX and transfected into SH-SY5Y cells. Cells were then serum starved overnight and treated with either nerve growth factor (NGF) and brain-derived neurotrophic factor (BDNF) at 37 °C for 0–12 h. Fifteen micrograms of total RNA were then collected from TrkA- and TrkB-SY5Y cells exposed to 0, 1.5, 4, and 12 h of NGF or BDNF and the gene expressions were profiled using the Affymetrix GeneChip 133A. Four and three replicates were performed for the TrkA and TrkB cells, respectively. The robust multiarray (RMA) procedure (14) was used to obtain the gene expression measures.

To perform network-based analysis of the data, we merged the gene expression data with the 33 KEGG regulatory pathways and identified 1,533 genes on the Hu133A chip that could be found in the 1668-node KEGG network of 33 pathways. Instead of considering all the genes on the Hu133A chip, we only focused our analysis on these 1,533 genes and aimed to identify which genes and which subnetworks of the KEGG network of 33 pathways are potentially related to the cell differentiation of TrkA-transfected cell lines. We analyzed the data using the HMRF model and obtained parameter estimates of $\alpha = -1.58$ and $\beta = 0.39$, indicating that there are more genes with similar expression patterns than those with different expression patterns. Our method identified 210 DE genes out of the 1,533 KEGG genes with posterior probability of being a DE gene greater than 0.5, among these 118 are connected on the KEGG pathways and 92 are isolated, not collecting to other DE genes. There is a large cluster of genes that are largely upregulated in the Trk A transfected cells but are downregulated in the Trk B transfected cells. Similarly, there is a cluster of genes that are upregulated in the Trk B-transfected cells but are downregulated in the Trk A transfected cells (see Fig. 2).

Among the 33 KEGG regulatory pathways, enrichment analysis using DAVID Tools (15) identified that the mitogen-activated protein kinase (MAPK) signaling pathway, focal adhesion pathway, and pathway related to prion diseases are enriched with p-values of 0.012, 0.029, and 0.05, respectively, of which the MAPK signaling pathway and the focal adhesion pathway are inter-connected. The MAPK (Erk1/2) signal transduction pathway is

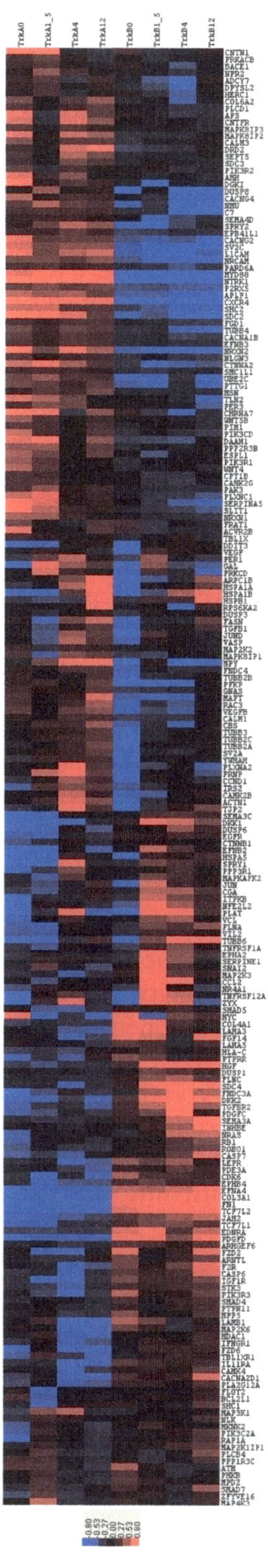

Fig. 2. Heatmap clustering plot of the 210 DE genes on the KEGG pathways, showing different expression patterns between the TrkA and TrkB time courses. The first four columns correspond to the TrkA time course experiments at times 0, 1.5, 4, and 12 h, the second four columns correspond to the TrkB time-course experiments at times 0, 1.5, 4, and 12 h.

expressed and active in both TrkA and TrkB expressing NB cells after specific ligand-mediated Trk receptor phosphorylation. The distinct role that this signaling pathway plays in the biologic heterogeneity of NB is not well known; however, we have shown that the time course of pathway activation by phosphorylation of signal effector proteins is different between TrkA- and TrkB-expressing NB cells, and this may, in part, explain the biological differences between TrkA- vs. TrkB-expressing tumors. To give a detailed comparison of TrkA- and TrkB-mediated genomic responses, we present in Figs. 3 and 4 the DE genes on the MAPK signaling pathway and on the KEGG focal adhesion pathway. On the MAPK pathway, it is not surprising that the TrkA/B shows different expression patterns. We also observed that a cluster of genes (or a subnetwork) in the neighborhood of ERK shows different expression patterns, including MEK2, MP1, PTP, MKP, Tau, cPLA2, MNK1/2, and c-Myc (see Fig. 3). This subnetwork, leading to cell proliferation and differentiation, may partially explain the difference in cell differentiation between the TrkA- and TrkB-infected NB cells. Another interesting subnetwork in the neighborhood of p38, including MKK3, MKK6, PTP, MKP, MAPKAPK, GADD153, and HSP27, also showed differential expression patterns. This subnetwork also related to cell proliferation and differentiation. Activation of these two subnetworks on the MAPK pathway may explain the different biological behaviors of these two types of NB cells, especially in terms of cell differentiation. MAPK signaling in the nervous system has been shown to promote a broad array of biologic activities including neuronal survival, differentiation, and plasticity. Regulating the duration of MAPK signaling is important in neurogenesis, and likely plays a similar role in the behavior of Trk-expressing neuroblastomas. Prolonged activation of MAPK is correlated with neurotrophin-dependent cell cycle arrest and terminal cellular differentiation in the PC12 pheochromocytoma cell line, whereas short-duration MAPK signaling is correlated with mitogenic and proliferative cell signaling in PC12 cells (15–19). TrkA-expressing NB cells treated with NGF (which activates MAPK) increase the number and length of extended neurites and decrease cell proliferation resulting in a more mature neuronal appearing cell, while TrkB-expressing NB cells treated with ligand (BDNF) increase cell proliferation without morphologic differentiation.

Increasing evidence suggests an important role for the focal adhesion kinase (FAK) pathway (see Fig. 4) in regulating cancer cell adhesion in response to extracellular forces or mechanical stress. Studies have demonstrated that tumor cells are able to regulate their own adhesion by over-expression or alteration in activity of elements within the FAK signaling pathway, which may have implications in the survival, motility, and adhesion of metastatic tumor cells (20). While mechanotransduced stimulation of the

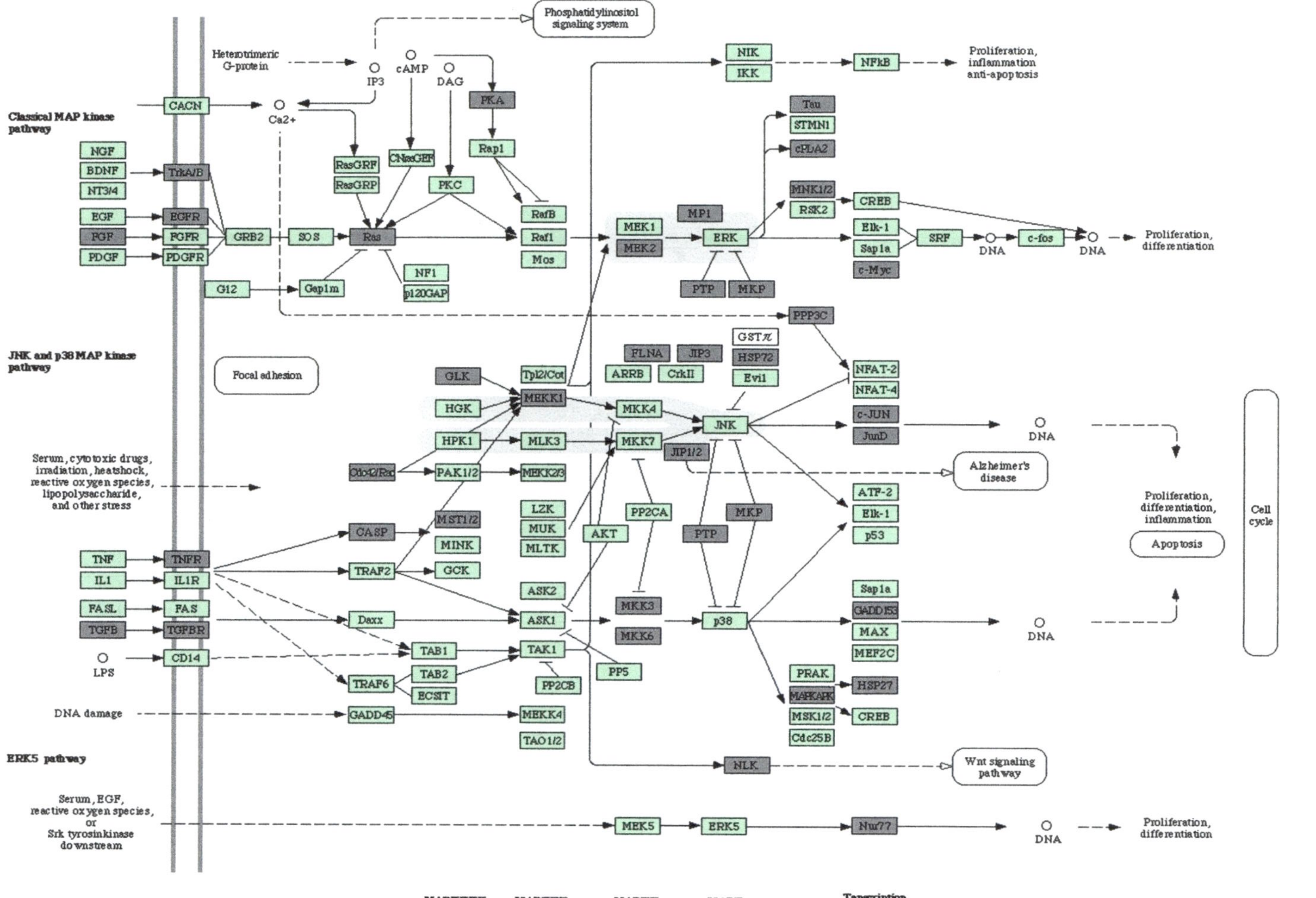

Fig. 3. Differential expression states for genes on the KEGG MAPK pathway, where genes colored in *dark gray* are multivariate differentially expressed and those colored in *light green* are equally expressed.

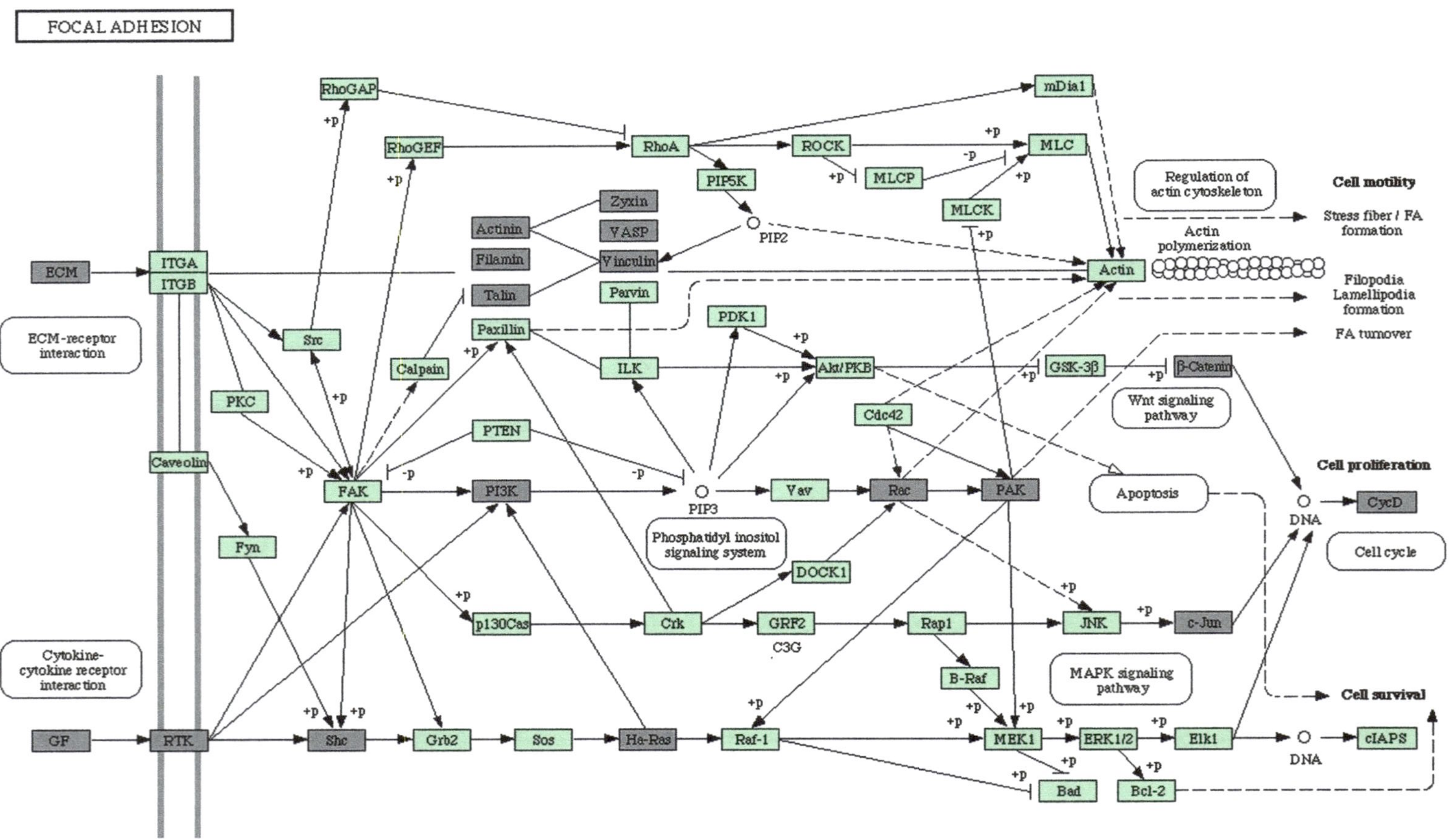

Fig. 4. Differential expression states for genes on the KEGG Focal Adhesion pathway, where genes colored in *dark* are multivariate differentially expressed and those colored in *light green* are equally expressed.

FAK signaling pathway appears to be a cell surface receptor independent process, the FAK pathway also acts downstream of receptor tyrosine kinases and has been shown to be phosphorylated in response to external cytokine/ligand stimuli. The insulin-like growth factor-1 receptor (IGF-1R) and FAK physically interact in pancreatic adenocarcinoma cells resulting in activation of a common signal transduction pathway that leads to increased cell proliferation and cell survival (21). In neuroblastoma, MYCN regulates FAK expression by directly binding to the FAK promoter, and increasing transcription of FAK mRNA. Beierle et al. (22) have correlated FAK mRNA abundance with MYCN expression in MYCN-amplified and non-amplified NB cell lines by real-time quantitative PCR, and their data suggest that MYCN regulation of FAK expression directly impacts cell survival and apoptosis. On the focal adhesive pathway, we observed a subnetwork of six genes, Actinin, Filamin, Talin, Zyxin, VASP, and Vinculin, that show differential expression patterns (see Fig. 4). In addition, PI3K and its neighboring genes GF, RTK, Shc, and Ha-Ras show differential expression patterns. We have not yet explored the regulation of FAK pathway activity by TrkA or TrkB expression and activation in our NB cell lines, but the differential expression states for genes on the KEGG FAK pathway suggest differential mediation by TrkA vs. TrkB, that may have downstream biological relevance.

Finally, on the pathway related to prion disease, we observed that Prion Protein (PrPc) and its neighboring genes HSPA5, APLP1, NRF2, and LAMB1 show differential expression patterns.

6. Conclusion and Discussion

In this paper we have proposed a hidden MRF model and an ICM algorithm that utilizes the gene regulatory network information to identify multivariate differentially expressed genes. The method extended the approach of (7) for univariate to multivariate gene expression data such as time course data. Also different from the approach of (6) for network-based analysis of microarray time-course gene expression data, this new approach identifies the genes that show different expression patterns over time rather than identifies the differentially expressed genes at each time point. Instead of assuming that all genes are independent as in the empirical Bayes approach of (4), our method models the dependency of the latent differential expression states based on the prior regulatory network structures. Simulation studies show that our methods outperform the methods that do not utilize network structure information. We applied our method to analyze the MTC data of TrkA- and TrkB-transfected neuroblastoma cell lines and identified the MAPK and focal adhesive pathways from the KEGG that are related to cell differentiation in TrkA-transfected cell lines. Note that the proposed methods can also be applied to other types of genomic data such as proteomic data and protein–protein interaction data.

In this paper, we analyzed the neuroblastoma MTC data using KEGG pathways and aimed to identify the KEGG pathways that may explain the differentiation states of the two different NB cell lines. However, the proposed methods can be applied to any other networks of pathways. If an investigator is only interested in a particular pathway, the proposed method can be applied to that particular pathway. If an investigator is interested in fully exploring his/her data and all available pathways, one should use a large collection of pathways, e.g., the pathways collected by Pathway Commons (http://www.pathwaycommons.org/pc/). It should also be noted that our proposed methods can include all the genes probed on microarray by simply adding isolated nodes to the graphs. A related issue is that our knowledge of pathways is not complete and can potentially include errors or misspecified edges on the networks. Although our simulations demonstrate that our methods are not too sensitive to the misspecification of the network structures, the effects of misspecification of the network on the results deserve further research. One possible solution to this problem is to first check the consistency of the pathway structure using the data available. For example, if the correlation in gene expression levels between two neighboring genes is very small, we may want to remove the edge from the pathway structure. Alternatively, one can build a set of new pathways using various data sources and compare these pathways with those in the pathway databases in order to identify the most plausible pathways for use in the proposed MRF method. For example, we can construct a large molecular network with the nodes being the gene products and the links extracted from the KEGG database, the Biomolecular Interaction Network Database (BIND) and Human Interactome Map (HIMAP) (see ref. (23)). This will provide more comprehensive description of known biological pathways and networks than using data from only one source.

In summary, generation of high-throughput genomic data together with intensive biomedical research has generated more and more reliable information about regulatory pathways and networks. It is very important to incorporate the network information into the analysis of genomic data in order to obtain more interpretable results in the context of known biological pathways. Such integration of genetic network information with high-throughput genomic data can potentially be useful for identifying the key molecular modules and subnetworks that are related to complex biological processes.

Acknowledgments

This research was supported by NIH grants R01-CA127334 and P01-CA097323. We thank Mr. Edmund Weisberg, MS at Penn CCEB for editorial assistance.

References

1. Lehmann KP, Phillips S, Sar M, Foster PMD, Gaido KW (2004) Dose-dependent alterations in gene expression and testosterone synthesis in the fetal testes of male rates exposed to Di (*n*-butyl) phthalate. Toxicol Sci 81:60–68
2. Seidel S, Stott W, Kan H, Sparrow B, Gollapudi B (2006) Gene expression dose–response of liver with a genotoxic and nongenotoxic carcinogen. Int J Toxicol 25:57–64
3. Yuan M, Kendziorski C (2006) Hidden Markov models for microarray time course data under multiple biological conditions (with discussion). J Am Stat Assoc 101(476): 1323–1340
4. Tai YC, Speed T (2006) A multivariate empirical Bayes statistic for replicated microarray time course data. Ann Stat 34:2387–2412
5. Hong FX, Li H (2006) Functional hierarchical models for identifying genes with different time-course expression profiles. Biometrics 62:534–544
6. Wei Z, Li H (2008) A hidden spatial–temporal Markov random field model for network-based analysis of time course gene expression data. Ann Appl Stat 2(1):408–429
7. Wei Z, Li H (2007) A Markov random field model for network-based analysis of genomic data. Bioinformatics 23:1537–1544
8. Wei P, Pan W (2008) Incorporating gene networks into statistical tests for genomic data via a spatially correlated mixture model. Bioinformatics 24:404–411
9. Besag J (1974) Spatial interaction and the statistical analysis of lattice systems. J R Stat Soc Ser B 36:192–225
10. Besag J (1986) On the statistical analysis of dirty pictures. J R Stat Soc Ser B 48:259–302
11. Aitchison J, Dunsmore IR (1975) Statistical prediction analysis. Cambridge University Press, London
12. Smyth GK (2004) Linear models and empirical Bayes methods for assessing differential expression in microarray experiments. Stat Appl Genet Mol Biol 3(1):Article 3
13. Brodeur GM (2003) Neuroblastoma: biological insights into a clinical enigma. Nat Rev - Cancer 3:203–216
14. Irizarry RA, Hobbs B, Collin F, Beazer-Barclay YD, Antonellis KJ, Scherf U, Speed TP (2003) Exploration, normalization, and summaries of high density oligonucleotide array probe level data. Biostatistics 4:249–264
15. Dennis G, Sherman BT, Hosack DA, Yang J, Gao W, Lane HC, Lempicki RA (2003) DAVID: database for annotation, visualization and integrated discovery. Gen Biol 4:P3
16. Tombes RM, Auer KL, Mikkelsen R, et al (1998) The mitogenactivated protein (MAP) kinase cascade can either stimulate or inhibit DNA synthesis in primary cultures of rat hepatocytes depending upon whether its activation is acute/phasic or chronic. J biochemistry 330(Pt 3):1451–1460
17. Kao S, Jaiswal RK, Kolch W, Landreth GE (2001) Identification of the mechanisms regulating the differential activation of the MAPK cascade by epidermal growth factor and nerve growth factor in PC12 cells. J Biol Chem 276(21):18169–18177
18. Marshall CJ (1995) Specificity of receptor tyrosine kinase signaling: transient versus sustained extracellular signal-regulated kinase activation. Cell 80(2):179–185
19. Qui MS, Green SH (1992) PC12 cell neuronal differentiation is associated with prolonged p21ras activity and consequent prolonged ERK activity. Neuron 9(4):705–717
20. Basson MD (2008) An intracellular signal pathway that regulates cancer cell adhesion in response to extracellular forces. Cancer Res 68(1):2–4
21. Liu W, Bloom DA, Cance WG, Kurenova EV, Golubovskaya VM, Hochwald SN (2008) FAK and IGF-IR interact to provide survival signals in human pancreatic adenocarcinoma cells. Carcinogenesis 29(6):1096–1107
22. Beierle EA, Trujillo A, Nagaram A, Kurenova EV et al (2007) N-MYC regulates focal adhesion kinase expression in human neuroblastoma. J Biol Chem 282(17):12503–12516
23. Alfarano C, Andrade CE, Anthony K, Hahroos N, Bajec M et al (2005) The biomolecular interaction network database and related tools 2005 update. Nucleic Acids Res 33:D418–D424

Chapter 9

Genomic Outlier Detection in High-Throughput Data Analysis

Debashis Ghosh

Abstract

In the analysis of high-throughput data, a very common goal is the detection of genes or of differential expression between two groups or classes. A recent finding from the scientific literature in prostate cancer demonstrates that by searching for a different pattern of differential expression, new candidate oncogenes might be found. In this chapter, we discuss the statistical problem, termed oncogene outlier detection, and discuss a variety of proposals to this problem. A statistical model in the multiclass situation is described; links with multiple testing concepts are established. Some new nonparametric procedures are described and compared to existing methods using simulation studies.

Key words: cDNA microarrays, Cancer, Differential Expression, Multiple Comparisons, Rank-based statistic

1. Introduction

Genomic technologies have permeated scientific research, especially in the area of cancer research (1). One of the major tasks in studies involving these technologies is to find genes that are differentially expressed between two experimental conditions, such as cancer and noncancer samples. This leads to a major multiple comparisons problem because of the dimensionality of the genes on the microarray. In practice, one has to perform thousands of tests of hypotheses. This has led to an explosion of literature on statistical methods for differential expression in genomic studies; see ref. (2) for a recent review on the topic. Many authors have advocated for control of the false discovery rate (FDR) (3) relative to the traditional familywise type I error (FWER). An opposing viewpoint on the issue has been given by Gordon et al. (4).

Less thought has been provided to the choice of test statistic to use. The two most commonly used statistics in the two-sample

Andrei Y. Yakovlev et al. (eds.), *Statistical Methods for Microarray Data Analysis: Methods and Protocols*, Methods in Molecular Biology, vol. 972, DOI 10.1007/978-1-60327-337-4_9, © Springer Science+Business Media New York 2013

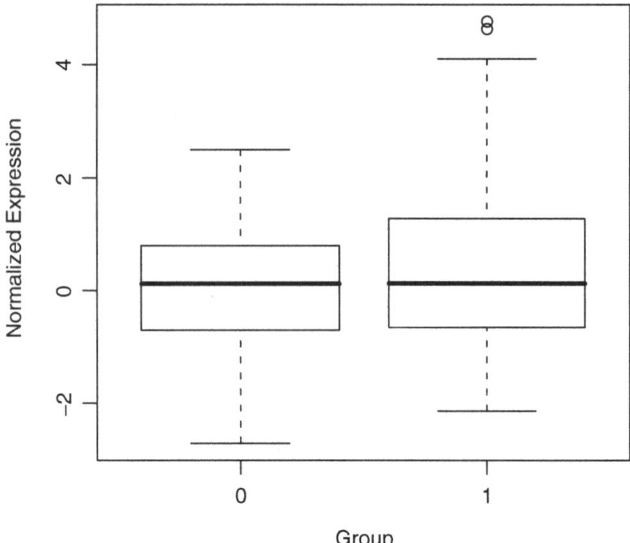

Fig. 1. Simulated data example.

problem are the *t*-test and the Wilcoxon rank sum test. Recently, Tomlins et al. (5) identified a gene fusion in prostate cancer. They discovered it by making the following observation. For certain genes, only a fraction of samples in one group were overexpressed relative to those in the other group; the remaining samples showed no evidence of differential expression. They used a method called Cancer Outlier Profile Analysis (COPA) to rank such genes using expression data generated by microarrays. Using such a score identified one of the genes involved in the fusion event. Tomlins et al. (5) proposed a statistic but gave no way of assessing its significance statistically. Recently, Tibshirani and Hastie (6), Wu (7), and Lian (8) have proposed modifications of two-sample *t*-tests to address the same problem.

What the COPA analysis reveals is that searching for differences in the centers of the distributions might not be sufficient in order to identify candidate oncogenes. An illustrative example is given in Fig. 1. The picture shows measurements from two populations, coded 0 and 1.

There are 50 observations in each group; the expression values are generated mixture of a standard normal random variable and a normal random variable with mean four in group 1. The mixing proportion for the distribution of measurements in group 1 is 0.1. Based on the boxplots, there is little difference in the centers of the distributions, but rather in the spread of the distribution and the fact that group 1 exhibits two outliers relative to group 0. Such a difference in distribution would not get picked up by the *t*-statistic; for the data in Fig. 1, the *p*-value from a two-sided *t*-statistic is 0.12. The difference is even less significant with a Wilcoxon rank sum statistic, which yields a *p*-value of 0.47. Alternatives to the *t*-statistic for two-group comparisons in microarray experiments have been suggested by Xiao et al. (9). In this chapter, we survey approaches to finding these cancer oncogenes, which we term outlier oncogenes. Note that while the term "outlier" typically has a pejorative meaning in statistics, from the point of view of cancer biology, such genes are of great interest. Under a model of tumorigenesis posited by Hanahan and Weinberg (10), there are several pathways by which cancer can develop:

1. Self-sufficiency in growth signals.
2. Insensitivity to antigrowth signals.
3. Evading apoptosis (cell death).
4. Limitless replicative potential.
5. Sustained angiogenesis.
6. Tissue invasion and metastasis.

If there are several different sets of genes by which such mechanisms can be activated, then it will be the case that a pattern in which a fraction of samples show differential expression for a given gene will be the case.

The chapter is structured as follows. In Subheading 2, we describe the data structure as well as a general statistical approach to identifying outlier genes. The idea is to treat outlier detection as a hypothesis testing problem with a data-driven null distribution. We then describe previous proposals in the literature for outlier detection in the two-group case; results for real and simulated data can be found in Ghosh and Chinnaiyan (11). Next, we extend the discussion to the multigroup case; this problem has only appeared to have been considered by Liu and Wu (12). We perform some simulation studies comparing the multigroup proposals. We conclude with some discussion in Subheading 3.

2. Methods

2.1. Data, Statistical Problem, and General Methodology

The data consist of Y_{gij}, where Y_{gij} is the gene expression measurement on the *g*th gene for the *i*th subject in group $g = 1,\ldots,G,\quad j = 0,\ldots,J$, $i = 1,\ldots,n_j$.

We will refer to the group with $j = 0$ as the baseline group. The notation $\mathbf{Y}_g$ denotes the gene expression profile of the *g*th gene across all subjects.

For a given gene *g*, we wish to identify the samples that have outlying values with respect to the gene. We assume that we will use the samples with $j = 0$ (i.e., the baseline group) as the baseline distribution with which to assess "outlyingness." We can state this as the following hypothesis testing problem: test the hypothesis H_0^i: the *i*th individual does not have outlying expression values of gene *g* relative

Table 1
Outcomes of n_1 tests of hypotheses regarding outlyingness

	Decide outlier	Decide non-outlier	Total
True non-outlier	U_1	U_3	n_{01}
True outlier	U_2	U_4	n_{11}
	R	Q	n

Note: The rows represent each sample being a true outlier or a true non-outlier. In the columns, decide outlier means that we reject H_0^i; decide non-outlier means that we fail to reject H_1^i. This is the results for a single gene; dependence on gene is suppressed in the notation

to the baseline population against H_1^i: the *i*th disease individual does have outlying expression values of gene g relative to the baseline population. For each gene we can perform such a test. Suppose that $J=1$; the result of the n_1 hypothesis tests is given in Table 1.

In Table 1 the vector (U_1, U_2, U_3, U_4) is unobserved, but the column totals and number of tests is known. This is because R and Q represent the number of hypotheses the data analyst decides to reject. Table 1 has a similar structure to many tables seen in the statistical literature on multiple testing (e.g., Table 1 of (3)). We will draw on the parallels between the two problems.

Based on the table, we can construct appropriate error measures to control. There are two kinds of errors that can be made. In Table 1, these correspond to U_2 and U_3. The random variable U_2 is the number of samples for which we declare them to be outlier when in truth they are not. The second error, represented by U_3 is to not declare a sample as an outlier when in fact it is an outlier. Most error rate measures control the first type of error. Two popular error measures to control are the familywise type I error (FWER) (13) and the false discovery rate (FDR) (3). In words, the FWER is the probability of making at least one false declaration of a sample being an outlier, while the FDR is the average number of false outliers among the samples declared to be outliers. Using the notation of Table 1, FWER equals $\Pr(U_1 \geq 1)$, while the FDR is $E[U_1 / R \mid R > 0]\Pr(R > 0)$.

We will use multiple testing ideas to develop outlier detection procedures. We first assume a single gene and treat the index g as fixed. We estimate F_{0g}, the distribution of gene expression of gene g in the baseline population. A simple estimator is the empirical cumulative distribution function:

$$\hat{F}_{0g}(y) = \sum_{i=1}^{n} {}_0 I\,(Y_{gi0} \leq y) / n_0.$$

Then we transform the gene expression measurements for the other samples by $\tilde{p}_{gi}^j = 1 - \hat{F}_{0g}(Y_{gij})$, $j = 1,\ldots,J$, $i = 1,\ldots,n_j$. We then use the $\tilde{p}_{gi}^j$ to test H_0^i versus H_1^i; in particular smaller values indicate evidence against H_0^i for the gth gene in group j. Note that if

F_{0g} were known, then $1 - F_{0g}(\Upsilon_{gij})$ would have a uniform (0, 1) distribution under H_0^i and a distribution of a stochastically smaller random variable under H_1^i. We will refer to the $\tilde{p}_{gi}^j$ as pseudo *p*-values. While these are not exactly *p*-values in the strictest sense because we are using a data-driven estimator of the null distribution $(\hat{F}_{0g})$, we will use these statistics as *p*-values.

Suppose we now wish to aggregate the results of the individual *p*-values across the samples. Three popular approaches are Bonferroni's method, Šidák's procedure, and the Benjamini–Hochberg method. Bonferroni's procedure declares observation *i* an outlier if $\tilde{p}_{gi}^j \le / n_j$. Šidák's procedure declares it to be an outlier if $\tilde{p}_{gi}^j \le 1 - (1 - \alpha)^{1/n_j}$. Finally the Benjamini–Hochberg procedure proceeds as follows:

1. Set an error rate α.
2. Sort the $\tilde{p}_{gi}^j$ in increasing order, $\tilde{p}_{g(1)}^j, \tilde{p}_{g(2)}^j, \ldots, \tilde{p}_{g(n_j)}^j$.
3. Take as outliers the samples corresponding to $\tilde{p}_{(g1)}^j, \ldots, \tilde{p}_{(g\hat{k})}^j$ where $\hat{k}_{jg} = \max\{1 \le i \le n_j : p_{g(i)}^j\} \le ia / n_j$. If no such $\hat{k}$ exits, conclude that there are no outliers.

Conditional on $\hat{F}_{0g}$, all the tests for outliers are independent. This implies that the Bonferroni and Šidák's procedure will control the FWER, while the Benjamini–Hochberg procedure will control the FDR exactly for the multiple outlier hypothesis tests for a fixed gene. Note that the multiplicity adjustment here is for the number of samples in the *j*th group.

Suppose we now wish to make gene-specific inferences regarding outlierness. This will be done on a genomewide scale. If we use the Bonferroni procedure, we compute

$$T_{jg}^B = \sum_{i=1}^{n_j} \Upsilon_{gji} I\{\tilde{p}_{gi}^j \le a / n_j\}. \tag{1}$$

For the Šidák procedure, it is

$$T_{jg}^S = \sum_{i=1}^{n_j} \Upsilon_{gji}\, I\{\tilde{p}_{gi}^j \le 1 - (1 - \alpha)^{n_j}\}, \tag{2}$$

while for the Benjamini–Hochberg procedure the score would be

$$T_{jg}^{BH} = \sum_{i=1}^{\hat{k}_{jg}} \Upsilon_{gji}, \tag{3}$$

defined in the previous paragraph. To derive the null distribution of (1), (2), and (3), we adopt a conditional approach in which we permute the labels of the *j*th group versus the baseline group. We now have a *J*-dimensional vector of *p*-values for the *g*th gene, $g = 1, \ldots, G$.

Suppose $J = 1$; the above methodology describes that which was proposed in Ghosh and Chinnaiyan (11). The multiplicity issue is now across genes, so we need to now make adjustments for G tests of hypotheses. We can then perform the usual multiple testing adjustments controlling either the FWER or the FDR. Based on the permutations, we can then adjust the p-values for multiple testing. A variety of procedures for doing so based on FWER can be found in Dudoit et al. (14). For presenting scientists with a list of genes calibrated in evidence for outlierness, we prefer the q-value approach of Storey and Tibshirani (15):

1. Order the G p-values as $p_{(1)} \leq \ldots \leq p_{(G)}$.
2. Construct a grid of L λ values, $0 < \lambda_1 < \ldots < \lambda_L < 1$ and calculate

   $\hat{\pi}_0(\lambda_1) = (\#\{p_j > \lambda\}) / (G(1-\lambda)), \quad l = 1, \ldots, L.$
3. Fit a cubic spline to $\{\lambda_l, \hat{\pi}_0(\lambda_1)\}$, $l = 1, \ldots, L$ and estimate π_0 by its interpolated value at $\lambda = 1$.
4. Estimate the FDR over a region $[0; t]$ $(0 < t \leq 1)$ by

$$\widehat{FDR}(t) = \frac{G\hat{\pi}_0(t)}{\hat{P}(R > 0)\sum_{i=1}^{G} I(p_i \leq t)},$$

   where $\hat{P}(R > 0) = \sum_i I(p_i \leq \alpha)$.
5. Calculate the q-value for the gth gene by

   $q(p_g) = \min_{t \geq p_g} \widehat{FDR}(t).$

In words, step (5) states the q-value is approximately the smallest false discovery rate at which we would reject the null hypothesis that there is no outlying expression for the gth gene in diseased relative to non-diseased samples.

2.2. Two-Group Problem: Previous Work

In this section, we describe previous proposals from the literature for the case of two groups, i.e., $J = 1$. The first proposal is that of Lyons-Weiler et al. (16), called the permutation percentile separability test (PPST). The statistic they propose is

$$\text{PPST}_g = \sum_{j=1}^{n_1} Y_{gi1} I\{Y_{gj1} > q_{95}(Y_{g1}, \ldots, Y_{gn_0})\}. \tag{4}$$

The COPA analysis approach of Tomlins et al. (5) calculates the following statistic for the gth gene:

$$\text{COPA}_g = q_r(\mathbf{Y}_g^*), \tag{5}$$

where $\mathbf{Y}_g^*$ is $\mathbf{Y}$ centered by the median and standardized by the median absolute deviation and $q_r(\mathbf{Y})$ denotes the rth percentile of the vector $\mathbf{Y}$. Tomlins et al. (5) take r to be 75, 90, and 99 but provide no measure of significance.

More recent approaches have been based on modifications of the two-sample t-tests. The first was by Tibshirani and Hastie (6). They use as their statistic

$$OS_g = \sum_{j=1}^{n_1} \Upsilon^{*}_{gj1} I\{\Upsilon^{*}_{gj1} > q_{75}(\mathbf{Y}^{*}_g) + \mathrm{IQR}(\mathbf{Y}^{*}_g)\}, \tag{6}$$

where OS stands for outlier sum, $\mathrm{IQR}(\mathbf{Y})$ denotes the interquartile range of a vector $\mathbf{Y}$. Tibshirani and Hastie (6) argue that the outlier sum score (6) is potentially more efficient than the *COPA* score because it sums over all outlying disease samples.

An alternative t-statistic has been given by Wu (7), which he terms the outlier robust t-statistic (*ORT*). The *ORT* differs from the *OS* method because the variation might be different between the two groups. The *ORT* statistic is given by

$$ORT_g = \frac{\sum_{j=1}^{n_1} \Upsilon_{gj1} I\{D_i = 1, \Upsilon^{*}_{gji} > q_{75}(\Upsilon_{g1}, \ldots, \Upsilon_{gn_0}) + \mathrm{IQR}(\Upsilon_{g1}, \ldots, \Upsilon_{gn_0})\},}{\mathrm{med}(\mathrm{med}_{1 \le n_0} \mid \Upsilon_{gi} - \mathrm{med}_{1 \le i \le n_0} \Upsilon_{gi} \mid, \mathrm{med}_{n_0+1 \le i \le n} \mid \Upsilon_{gi} - \mathrm{med}_{n_0+1 \le i \le n} \Upsilon_{gi} \mid)}, \tag{7}$$

where $\mathrm{med}(\mathbf{v})$ denotes median of a vector $\mathbf{v}$.

Finally, there is the proposal of Lian (8). The issue that he addresses in his work is that the number of outlying samples is unknown. Given that, what one should try to do is pool the non-outlying samples from group 1 into group 0 and to then calculate a modified t-test based on the "new" groups. Formally, he considers a statistic in which one calculates a modified t-statistic over all possible allocations of the sorted gene expression values from group 1 to group 0. The statistic of interest is then a maximum over all possible partitions. In analyzing the work of Lian (8), we find that there is another level of multiplicity, that of searching over all possible partitions. This is not addressed explicitly in the work of Lian (8). However, it is possible that the results of Benjamini and Heller (17) might be of use here. We leave this as an open conjecture.

It helps to better situate these proposals in the context of a proper statistical model. Ghosh and Chinnaiyan (11) describe a mixture model for outlier gene detection:

$$\Upsilon_{gi} \mid Z_i = 0 \overset{\mathrm{ind}}{\sim} F_{0g}(y),$$

$$\Upsilon_{gi} \mid Z_i = 1 \overset{\mathrm{ind}}{\sim} \pi_{0g} F_{0g}(y) + (1 - \pi_{0g}) F_{1,gi}(y), \tag{8}$$

$$\pi_{01}, \ldots, \pi_{0G} \overset{\mathrm{iid}}{\sim} p\delta_1 + (1 - \delta) F_p(\pi), \tag{9}$$

where F_{0g} and F_{1gi} are distribution functions. If π_{0g} comes from the first mixture component, then there is no differential expression for the gth gene. To make the model sensible, we need that smaller values of π_{0g} correspond to an increased likelihood of coming from the distribution function F_p. What all the previous authors propose is testing H_0: $\pi_{0g} = 1$ for $g = 1,\ldots,G$.

If we were to only consider the first two equations in (8), then it is also easily seen that we are able to determine which samples are outliers relative to the baseline distribution for a given gene. While the hypothesis described in the previous paragraph is being tested genomewide, the hypothesis being described here occurs is a sample-specific one. Thus, if investigators have one gene of particular interest, they could then determine which samples show outlying values of expression.

In Ghosh and Chinnaiyan (11), several simulation studies are performed to determine the relative performance of the proposed methodologies. The proposal of Ghosh and Chinnaiyan corresponds to the work described in the previous section for $J = 1$. Roughly speaking, what they found was that when the data were close to normality, then the modified *t*-test proposals worked the best. However in other cases where there is clear evidence of non-Gaussianity, the nonparametric methods described in Subheading 2 performed better.

2.3. Multivariate Case: More Than Two Groups

In thinking about comparing more than two groups for outlier differential expression, we need to consider vector-valued versions of the pseudo *p*-values. Consideration of multivariate *p*-values has received much less attention in the statistical literature. Ploner et al. (18) propose a multivariate density estimation for the local false discovery rate of Efron et al. (19), while Chi (20) proposes multivariate extensions of the Benjamini–Hochberg procedure. We use the second approach here based on biological considerations, as will be discussed shortly.

We begin by stating the mixture model for the multivariate *p*-values. Let $\mathbf{p}_1,\ldots,\mathbf{p}_G$ generically denote $J \times 1$ vectors of *p*-values. We adopt the random effects model specified in Chi (20): for $i = 1,\ldots,G$,

$$(\mathbf{p}_i, Z_i)\tilde{\ } \text{iid}(\mathbf{p}, Z),$$

$$Z_i \sim \text{iid Bernoulli}(\pi_0),$$

$$p_{ji} \mid Z_i = 0 \sim \text{iid}\, U(0,1), \tag{10}$$

$$\mathbf{p}_i \mid Z_i = 1 - \mathcal{P}.$$

For this model, we assume that for each gene, there exists an unknown state variable Z that is Bernoulli distributed with probability π_0. In this model, $Z = 1$ denotes that the gene comes from the population of outlier genes, while $Z = 0$ represents that the gene is not from the population of outlier genes. For the non-outlier genes, we assume that the components of the multivariate *p*-value are independent and identically distributed from the uniform (0, 1) distribution; this is given in (10). Under the alternative (i.e., the population of outlier genes), we assume that the joint distribution has a distribution *P*, as evidenced in (9).

Chi (20) points out the fact that (10) is a crucial assumption in the model. For our setting, if we condition on the distribution of the gene expression values in the baseline group, then all the components of $\mathbf{p}_i$ are independent since different samples are used for calculating the *p*-value in the *j*th group, $j = 1,\ldots,J$. Thus, (10) is a reasonable assumption here.

Given that the marginal distribution of $\mathbf{p}_1,\ldots,\mathbf{p}_G$ follows a mixture model, there is an inherent identifiability issue regarding π_0 , the proportion of genes for which $Z=1$. Here and in the sequel, we assume that the parameter is identifiable, sufficient assumptions for which are given in Subheading 3.1 of Genovese and Wasserman (21).

Setting the false discovery rate to be η, an algorithm describing application of the Benjamini–Hochberg procedure to multivariate *p*-values is summarized by (3.5) of Chi (20) and is summarized as follows:

1. Construct a function *R* with the property that
   $D_t = \{\mathbf{p} \in [0,1]^J : R(\mathbf{p}) \le t\}$

has Lebesgue measure *t*.

2. Apply the Benjamini–Hochberg procedure to $s_i = R(\mathbf{p}_i)$, $i = 1,\ldots,G$;

   Sort s_i in increasing order, $s(1) \le \ldots \le s(G)$.

   Define $\hat{k} = \max\{1 \le i \le G : s_{(i)} \le i\eta / G\}$.

Conclude that genes with scores $s_{(1)},\ldots,s_{(k)}$ are from the outlier gene population, while the rest are not. If there exists no such $\hat{k}$, the conclude that there are no outlier genes.

The reason for the constraint on the function *R* is because we need under the null hypothesis that $R(\mathbf{p}) \sim U[0,1]$. The important question then becomes what function *R* to use in step 1 of the algorithm. Here biological considerations guide us. If we expect gene expression to show outlying values for only one group of samples, then a natural function to use is $R_1(\mathbf{p}) = \min p_j$. This is that the expression pattern of a gene exists in a “biological subtype” of an existing class. If instead we expect outlying values to be consistent across all classes of samples, then a natural function to use is $R_2(\mathbf{p}) = \prod p_j$. Such a function would be in line with a model of tumorigenesis *n* which there is outlying expression across several groups. These are in fact the functions described in Example 3.1 of Chi (20).

Given that we are applying the Benjamini–Hochberg method to the multivariate *p*-values, it will control the false discovery rate at level α under the model described above. Note that the values of s_i under the null hypothesis will be independent and identically distributed from a *Uniform*(0, 1) distribution.

More generally, the procedure will still control the false discovery rate if the s_i values for the genes corresponding to the null distribution (i.e., non-outlying distribution) satisfy a so-called positive regression dependency condition described in Benjamini and Yekutieli (22).

It is not apparent how to modify the approach of Ploner et al. (18) to adapt to these two biological hypotheses. In the work of Liu and Wu (12), the authors focus on the second scenario but not on the first. In addition, their approaches are based on modifications of *F*-statistics that are analogous to the modified *t*-statistic proposals given in the previous section. The methodology being proposed here is more nonparametric in nature. We expect that the proposed methodology would be less efficient than that of Liu and Wu in the case of Gaussian data but that the proposed methodology would be more powerful in non-Gaussian cases as well as when there exist subtype-specific outlier genes. We test that next using simulation studies.

2.4. Simulation Studies

In this section, we report on the results of simulation studies for the various multigroup proposals being considered. In particular, we simulated data on four groups, each with 15 samples; 1,000 gene measurements were considered. The first group is treated as the baseline group. We assumed two scenarios:

1. One oncogene was expressed in all three groups relative to baseline.
2. One oncogene was expressed in one group relative to the baseline distribution.

We considered situations in which there was overexpression of five samples. Two distributional situations as well. First, we assumed that the expression of the non-oncogenic genes were distributed as a standard normal random variable. The gene expression of the oncogene was simulated as a normal distribution with mean two and variance one. The second scenario used an exponential distribution with baseline and overexpressed distributions having mean values 1 and 3. We compared the following methods: the ordinary *F*-statistic, the two outlier *F*-statistics proposed by Liu and Wu (12), and the outlier detection procedures proposed here with R_1 and R_2 as combining functions and the Bonferroni and Benjamini–Hochberg procedures, described in Subheading 2. We used R_1 for scenario 1 and R_2 for scenario 3. The Šidák method gave identical results to the Bonferroni, so we do not consider that procedure here.

The receiver operating characteristic curves are given in Figs. 2 and 3. Based on that, we find that the first outlier *F*-statistic proposed by Liu and Wu (12), along with the Bonferroni method with appropriate combining function, appears to yield the best performance

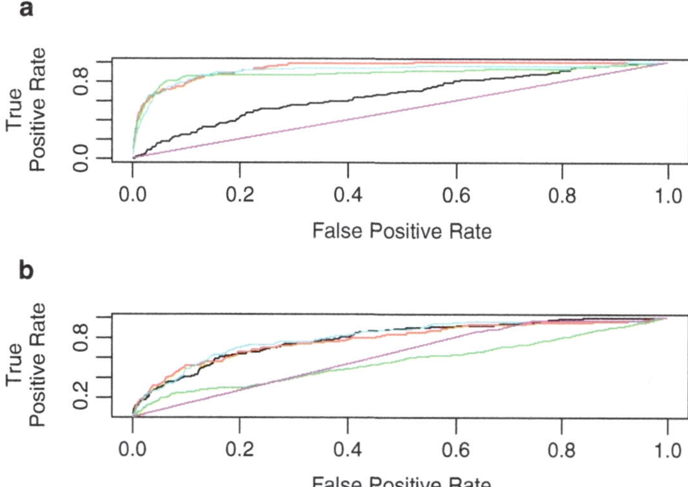

Fig. 2. Comparison of receiver operator characteristic (ROC) curves for denotes the usual *F*-statistic; the *red* and *green lines*, respectively, represent the first and second proposed outlier robust *F*-statistics of Liu and Wu (12). The *turquoise* and *purple lines* represent the Bonferroni and Benjamini–Hochberg methods. (**a**) Represents situation where one oncogene is overexpressed consistently in three groups, while Figure (**b**) represents situation where one oncogene is overexpressed only in one group.

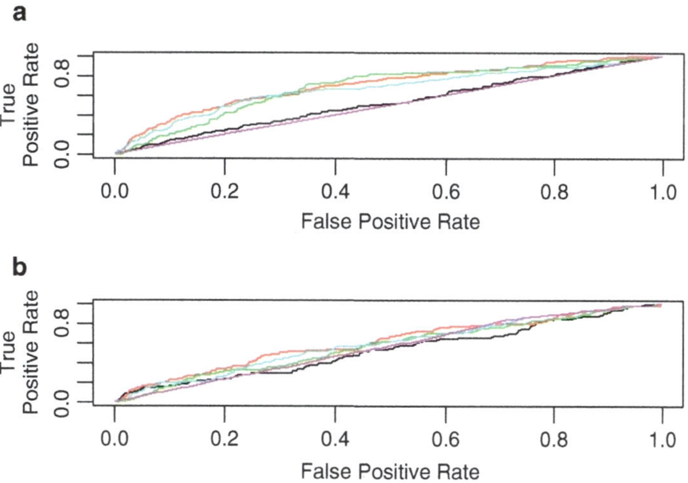

Fig. 3. Comparison of receiver operator characteristic (ROC) curves for proposed methods in exponentially distributed situation. See Fig. 2 for plot and line color definitions.

characteristics. It should be noted that for the second scenario (the oncogene being expressed differentially in only one group), all methods tend to perform fairly poorly.

3. Conclusion

In this section, we have described several methods for the identification of outlier oncogene detection using genomic data. While much of the discussion has used gene expression data, the methodology is fairly general and could potentially apply to high-throughput data from other platforms, such as copy number microarrays or tiling microarrays.

Much of the methodology discussed here has focused on assessing the marginal distributions of individual genes to identify candidate outlier oncogenes. Another observation made in passing by Tomlins et al. (5) was that the ERG-ETV1 gene pair showed evidence of mutually exclusive expression, i.e., if the expression of the gene in one sample was high, then that of the other gene was off and vice versa. This suggests an approach in which one searches for gene pairs that show evidence of a particular differential expression. While there is an expansive literature on finding differentially expressed genes, that of identifying differentially expressed gene combinations is much less, with some notable exceptions (23, 24). MacDonald and Ghosh (25) developed a thresholding-based algorithm for identifying candidate outlier oncogene pairs, but further work is definitely needed here. Another link that has been brought up here is the link between outlier detection with multiple testing. This provides a natural connection to the multiple testing literature. However, further work is needed here as well, as the multiclass outlier group detection treated in Subheading 2.3. Development of multivariate multiple testing procedures is currently under investigation.

Acknowledgments

The author would like to acknowledge the support of the Huck Institutes of Life Sciences at Penn State University and NIH Grant R01GM72007.

References

1. Ludwig JA, Weinstein JN (2005) Biomarkers in cancer staging, prognosis and treatment. Nat Rev Cancer 11:845–856
2. Ge Y, Dudoit S, Speed TP (2003) Resampling-based multiple testing for microarray data analysis. Test 12:1–44
3. Benjamini Y, Hochberg Y (1995) Controlling the false discovery rate: a practical and powerful approach to multiple testing. J R Stat Soc Ser B 57:289–300
4. Gordon A, Glazko G, Qiu X, Yakovlev A (2007) Control of the mean number of false discoveries, Bonferroni and stability of multiple testing. Ann Appl Stat 1:179–190
5. Tomlins SA, Rhodes DR, Perner S, Dhanasekaran SM et al (2005) Recurrent fusion of TMPRSS2

and ETS transcription factor genes in prostate cancer. Science 310:644–648

6. Tibshirani R, Hastie T (2007) Outlier sums for differential gene expression analysis. Biostatistics 8:2–8
7. Wu B (2007) Cancer outlier differential gene expression detection. Biostatistics 8:566–575
8. Lian H (2008) MOST: detecting cancer differential gene expression. Biostatistics 9:411–818
9. Xiao Y, Gordon A, Yakovlev A (2006) The L_1-version of the Cramer-von Mises test for two-sample comparisons in microarray data analysis. EURASIP J Bioinform Syst Biol 85769
10. Hanahan D, Weinberg RA (2000) The hallmarks of cancer. Cell 100:57–70
11. Ghosh D, Chinnaiyan AM (2008) Genomic outlier prole analysis: mixture models, null hypotheses and nonparametric estimation. Biostatistics (Advance Access published on June 6, 2008). doi:10.1093/biostatistics/kxn015
12. Liu F, Wu B (2007) Multi-group cancer outlier differential gene expression detection. Comput Biol Chem 31:65–71
13. Shaffer J (1995) Multiple hypothesis testing. Annu Rev Psychol 46:561–584
14. Dudoit S, Yang YH, Callow MJ, Speed TP (2002) Statistical methods for identifying differentially expressed genes in replicated cDNA microarray experiments. Statistica Sinica 12: 111–140
15. Storey JD, Tibshirani R (2003) Statistical significance for genomewide studies. Proc Natl Acad Sci USA 100:9440–9445
16. Lyons-Weiler J, Patel S, Becich MJ, Godfrey TE (2004) Tests for finding complex patterns of differential expression in cancers: towards individualized medicine. BMC Bioinform 125:110
17. Benjamini Y, Heller R (2008) Screening for partial conjunction hypotheses. Biometrics (Published online February 6, 2008). doi:10.1111/j.1541-0420.2007.00983.x
18. Ploner A, Calza S, Gusnanto A, Pawitan Y (2006) Multidimensional local false discovery rate for microarray studies. Bioinformatics 22:556–565
19. Efron B, Tibshirani R, Storey JD, Tusher V (2001) Empirical Bayes analysis of a microarray experiment. J Am Stat Assoc 96:1151–1160
20. Chi Z (2008) False discovery control with multivariate *p*-values. Electron J Stat 2:368–411
21. Genovese CR, Wasserman L (2004) A stochastic process approach to false discovery control. Ann Stat 35:1035–1061
22. Benjamini Y, Yekutieli D (2001) False discovery control under dependency. Ann Stat 29:1165–1188
23. Dettling M, Gabrielson E, Parmigiani G (2005) Searching for differentially expressed gene combinations. Genome Biol 6:R88
24. Xiao Y, Frisina R, Gordon A, Klebanov L, Yakovlev A (2004) Multivariate search for differentially expressed gene combinations. BMC Bioinform 26:164
25. MacDonald JW, Ghosh D (2006) COPA-cancer outlier prole analysis. Bioinformatics 22:2950–2951

Chapter 10

Impact of Experimental Noise and Annotation Imprecision on Data Quality in Microarray Experiments

Andreas Scherer, Manhong Dai, and Fan Meng

Abstract

Data quality is intrinsically influenced by design, technical, and analytical parameters. Quality parameters have not yet been well defined for gene expression analysis by microarrays, though ad interim, following recommended good experimental practice guidelines should ensure generation of reliable and reproducible data. Here we summarize essential practical recommendations for experimental design, technical considerations, feature annotation issues, and standardization efforts.

Key words: Data quality, Experimental design, Quality parameters, Annotation, Standardization

1. Introduction

Since the first reports on genome-wide monitoring of abundance of RNA species in biological specimen were published in the mid-1990s (1–3), use of microarrays or beadarrays (depending on provider), the number of publications as well as number of patents has increased almost exponentially (4). A major point of concern which is still under discussion to date is how to achieve good quality data and how to improve the comparability of data from laboratory to laboratory and from platform to platform. The lack of data comparability and reproducibility of data has led many to distrust the technology per se. On the technical side, there are plenty of recommendations (e.g., (5, 6)) and rule of thumbs, but they leave lots of room for subjective interpretation and use. Divergent quality assessments in sampling, processing, analysis, and reporting lead to heterogeneous data quality which is often not traceable by external researchers. Consequently, data from two laboratories have been rarely comparable. As lack of comparability leads instantly to a lack of credibility, researchers, industry, and providers of platforms are working on standardization guidelines for microarray experiments

Andrei Y. Yakovlev et al. (eds.), *Statistical Methods for Microarray Data Analysis: Methods and Protocols*, Methods in Molecular Biology, vol. 972, DOI 10.1007/978-1-60327-337-4_10, © Springer Science+Business Media New York 2013

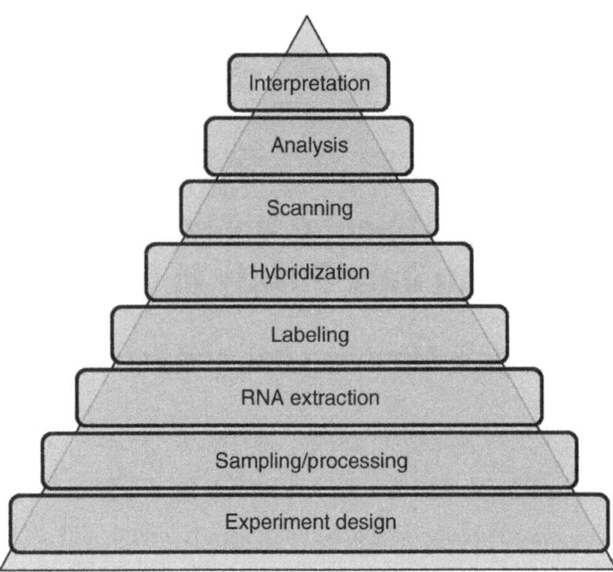

Fig. 1. The individual steps of microarray data generation. Each step should be subjected to thorough quality control.

and data sharing to assure that quality parameters can be met. It has become evident over the past years that microarray data are indeed comparable and reliable, as long as standard procedures and guidelines are followed strictly. In 2006, the FDA-initiative "MAQC" ("microarray quality control") published a series of reports on their analysis of public datasets, showing that microarray data from widely used platforms in a multitude of laboratories were comparable, because best practice guidelines were followed (7). Meanwhile, several guidelines have been published which provide guidelines and propose best practices in microarray experiments (e.g., (6, 8, 9)).

According to the definition by the OECD (Organization for Economic Co-operation and Development) Glossary of Statistical Terms (http://stats.oecd.org), high quality of data can be characterized as being relevant, accurate, credible, timely, accessible, interpretable, and coherent. Thus, as we will exemplify for microarray studies, achievement of data quality is a multifaceted problem, and data quality becomes a concept. Microarray data generation and analysis is an extremely complex process with many steps involved which require expert experience of the personnel with diverse background and specialized skills performing those steps. Each of those steps (Fig. 1) requires stringent quality control assessment.

To understand the necessity of standardization for data reliability, we will first touch the issues of sources of variation in a microarray experiment and the influence of experimental design, and sampling as essential cornerstones for data quality, and then focus on microarray batch processing and analysis. We will pay special attention to an aspect which has so far been largely undermined in the discussions, the issue of annotation quality of probes or probe sets on microarray platforms which has a huge impact on the interpretability of the data. Furthermore we will reflect on the current possibilities of data standardization and the repositories for accessibility to the scientific community.

2. Technical Aspects of Data Quality

2.1. Experimental Design: Avoidance of Bias

Microarray data generation always begins with a planning phase, the experimental design, never at a later phase. It is useful to know before the first sample is processed what the goal of the study is and how to get there, which go/no-go criteria to set and what the analysis approach will be. Each step along the process needs to be considered in this early phase, as flaws along the way may have detrimental impact on the interpretation of the results. It is useful for go/no-go-decisions, to set thresholds which allow you an easy decision for moving on to the next step in the experiment when the criteria are met, or move back, if they are not met (if moving back is possible). As we shall discuss below, it is almost impossible to provide objective quality parameter for any step of the experiment; most of them are based on experience or are recommendations with usually a large margin and still room for subjectivity.

Measurements taken during an experiment underlie the influence of fluctuation and instability of technical factors. Hence measurements tend to deviate from the "true" value, which is called experimental error. The degree of this error may be influenced by variations in time of day, temperature, cell-culture medium constitution, or even minor differences in manual handling. Many other technical sources of random variation can be envisioned.

In addition, biological variation which is inherent to living beings is a source of measurement variation. Such biological variation is observable between species, within a species, and even within an individual. Design strategies address technical and biological variation by repeating measurements of the same sample.

Experimental design addresses random and systematic variation by application of strategies of randomization, blocking, and replication, by increasing the number of samples per study group, by stratification of samples according to their demographic origin, by automation and standardization, and other design principles (10). Blocking refers to grouping experimental units, for example, patient samples, into homogeneous sets with respect to a factor or factors believed to be responsible for uncontrolled variability, and then assuring that all combinations of the factors of interest are included in each block. Randomization helps alleviate biases due to uncontrolled factors. A random sample is one in which each individual of a study population or a sample has an equal chance of being selected.

The ultimate goal is the increase of both accuracy and reliability of data, optimize relevance, timeliness, and information content of a study. Its purpose is to establish a cause–effect relationship under systematic variation of experimental conditions. Along the experiment, observations are collected and the design of an experiment essentially determines the pattern in which these observations are collected. Systematic variation of a measurement value from its true value, called bias, can be introduced by incorrect instrument settings, impurities in a production batch, differences in handling between laboratory personnel, and many other reasons. Microarray experiments can also be influenced by bias, as, e.g., scanner settings can be different from part of an experiment to another and thus result in systematic variation of measurements.

Experimental design also determines handling and processing of samples. Where the number of samples exceeds the number of samples that can be handled by one person at a time, block-design needs to be applied to avoid bias by handling differences. Where more than one person works on the samples, the design has to take care that not a single person works on samples of one analysis-group, while the other person processes the other (which would induce large handling-bias effect, fogging the biological readout). Experimental design ensures that only one reagent lot is used, and, where this is not possible, that the different lots are allocated to the samples in a design which eliminates lot-effects. This list could be extended drastically and it becomes obvious, that the effort which needs to be put into experimental design is immense, if the goal is to resolve biological questions without the negative impact of technical noise or bias.

2.1.1. Bias

The introduction of bias in microarray experiments may occur at different levels, but two kinds of bias may be of particular interest. Selection bias refers to which subjects, which experimental units are included in a study, other bias may occur at the measurement (i.e. observational) and interpretation level.

2.1.2. Selection Bias

The goal of sample selection is to achieve a sample population which is representative of the entire target population. Failure in doing so, for example in studies with small sample number, may introduce bias which results in study data which may explain the biological situation in this specific population and reflects rather random sampling noise from sample to sample, but which by no means should be extrapolated to the entire population which is targeted. This ultimately leads to an overestimation of the sensitivity of the assay. Where only subjects with a certain (disease) history or other features volunteer to participate in a study, or where self-assessment of the patients is an inclusion-criteria, there is a high risk for participation bias. Similarly, a high risk of selection bias exists for example in self-selection studies where only a specific subgroup of patients volunteers to participate. Those who volunteer may be healthier or more health-conscious than those who refuse, leading to an over-assumption of the efficacy. On the other hand, patients dropping out from a study may be different from patients who remain in the study; consequently, the exclusion of those patients from the analysis will introduce a selection bias. Control groups should be matched as stringent as possible to the groups of interest. For instance, young, disease-free subjects should be compared to non-diseased, but age-matched patients (11). Disregarding age as a conceptual factor in this example may introduce a bias towards biological factors which may interfere with the target observation.

2.1.3. Observational Bias

Measurement bias or observational bias occurs quite frequently in preclinical as well as clinical studies. Especially in long-term clinical trials, methods of data collection may differ slightly (e.g., over time, or from center to center, or from sample group to sample group). This is especially critical when different methods are applied to different experimental groups. For instance, only patients with negative test results and those with benign-appearing lesions undergo a reference test. The tendency that patients with positive test results undergo more intensive follow-up investigations may lead to nonrandomly distributed missing data between diseased and disease-free subjects.

2.1.4. Bias at Specimen/Tissue Collection and Processing

Several sources of variation with risk of bias should be considered when planning and performing genomic and proteomics studies. As mentioned earlier, the concept of randomization should be applied to the specimen sampling process to ensure equal conditions across the experimental units. In the setting of toxicological studies, vehicle treated animals may be sacrificed first before treated animals to avoid sample carryover from samples of treated to nontreated animals. As these procedures may be time-consuming, the time between last dose and tissue harvest can be much longer for the treated animals than for the vehicle-treated ones. Consequently, the RNA response to the drug and other biological responses to the environment (e.g., circadian rhythm, time-lapse from food-intake to harvest) are systematically different for the longer-surviving animals. Additionally, another collection bias relates to the time from food intake to specimen collection which should be blocked for, or randomized. When designing a sample collection scheme, circadian and seasonal should be accounted for (12) and stress exposure of the patients and animals should be balanced across treatments.

One has to pay extra attention that specimen are treated, stored, conserved, and processed equally across experimental groups. Variation caused by differences in sample handling may become of critical importance in the data interpretation, if not taken care of. Such differences may occur at the level of method of harvest, storage duration and temperature, and whether tissue samples are conserved in paraffin blocks or as snap frozen samples, different use of mRNA extraction or other protocols. Standardization of protocols is essential. Randomization and blocking of samples among operators should be applied to avoid bias and batch effects. As systematic variation can be introduced by the use of different lots of reagents or batches of microarrays, meticulous recording of the experimental conditions (temperature of hybridization, lot-number, operator, etc.) is essential and will be useful in the analysis of the data (13–15). Consequently, the entire process should be standardized as much as possible and any deviation from the protocol has to be recorded. Obviously, technical variation of the measurement process can be controlled for to a certain extent.

2.2. Sample and RNA Quality

A number of methods for RNA preparation have been tested and reviewed elsewhere. RNA integrity is of utmost importance when measuring abundance of RNA specimen with microarray technology (16). Any contamination with genomic DNA, protein, or chemicals can be harmful for the downstream steps in the protocol, such as reverse transcription (RT) to cDNA. Contaminating agents may lead to an insufficient RT reaction, a decrease in cDNA yield which may become too low for the labeling reaction, and cDNA-species which are too short for hybridization to the array.

A major threat to RNA integrity is the omnipresence of RNA-digesting RNase enzymes. The degree of RNA degradation is an important quality check point (16). A standard rule of thumb for acceptable RNA purity is that the ratio of optical density of 260 and 280 nm, which can be measured in spectrophotometer, should be around 2.0. The traditional interpretation of RNA quality, however, differed from laboratory to laboratory and from person to person, such that thresholds could be changed slightly. Standardization and an automated process for the evaluation of RNA quality have replaced the subjective method. These "microfluidics" platforms, such as the 2100 Bioanalyzer by Agilent, generate electropherograms of minute amounts of RNA in an automated fashion by size-separation of RNA and quantitation and quality assessment. Intact total RNA yields two sharp peaks of 28S RNA and 18S RNA. In addition rRNA content can be measured. Originally, a rule of thumb stated that good quality total RNA has a 28S/18S ratio should be around 2.0. Dumur et al. found that it is actually the percentage of rRNA which is present in the total RNA mixture, which is more relevant (17). Despite 28S/18S ratios of well below the standard cutoff of 2.0, a percentage of 30% rRNA yielded good quality array data (as indicated by median lengths of cDNA and cRNA products of 2.0–3.0 kb). Recently, an algorithm has been developed to be used in combination with microfluidics-workstations, generates RIN (RNA integrity number) to each sample and helps assess the quality of the RNA samples in a more objective fashion (18). The RIN is a numerical value on a scale from 1 (totally degraded) to 10 (intact). So far there is now general recommendation for RIN cutoff value for microarray experiments.

3. Experimental "Standards"

As we have seen in the paragraphs above, standardization of protocols within a site is a detrimental tool to alleviate nonrandom data variation. Standardization is even more critical when data from multicenter clinical studies need to be incorporated in a single analysis or need to be compared, or when we consider the future perspective of personalized medicine when an individual patient's genotype or expression profile could be used as diagnostic or prognostic readout. Only the standardization of procedures ensures that data are reliable, comparable, and interpretable. The call for standardization and quality control across laboratories and centers and quality control is heard, but final decisions on a consensus are still needed.

The lack of commonly agreed quality standards had early on sparked discussions on the reliability of the technology. Validity of microarray technology was disputed in publications and comments like "Microarrays and molecular research: noise discovery?" (19) "An Array of Problems" (20), "Arrays of Hope" (21) and "In Praise of Arrays" (22). Publications expressing concerns about the reproducibility of microarray data (23–26) were countered by those showing increased reproducibility (27–29).

Thus far, no commonly applicable best practice guidelines have been established for the most widely used platforms, spotted cDNA-array (Agilent), spotted oligonucleotides (Agilent), native oligonucleotide arrays (Affymetrix), and most recently the bead-system by Illumina. The increasing number of genomics data submission to regulatory agencies has made it critical for researchers in academia and industry to observe minimal quality standards and introduce best practice standards. However, those standards still vary from laboratory to laboratory.

Quality assessment of microarrays is highly recommended. Accuracy and precision are the two features which determine the quality of a study. Accuracy describes sensitivity and specificity and is a measure for how close an observation lies to the true value. To assess accuracy calibration samples of known concentration can be used (30, 31). Accuracy not only is an assessment of the array but also of the sample quality.

Precision describes how well the measurement can be reproduced and is usually an average replication error. Only the combination of accuracy and precision give a good estimate for the quality of the study. Having high precision but low accuracy indicates good reproducibility of the data, but the data do not reflect the true values. In contrast, highly accurate data may not be reproducible, which reveals problems along procedure.

The question of whether microarray data are accurate and reproducible has been long discussed (32, 33). To address the pending issue, the MAQC study assessed the performance of five different microarray platforms (Applied Biosystems, Affymetrix, Agilent one-colour, GE Healthcare, Illumina) on two commercially available RNA samples which were processed in three laboratories (7). The result shows that intra- and inter-laboratory comparisons as well as inter-platform comparisons.

In a milestone study, Shi et al. showed that standardization of procedures indeed leads to increased comparability of microarray data. The MicroArray Quality Control Consortium (MAQC) with 137 participants from 51 organizations (7) generated and analyzed a dataset consisting of two commercially available RNA samples of high-quality Universal Human Reference RNA (UHRR and Human Brain Reference RNA) which were mixed in four titration pools. Expression data were measured on seven microarray platforms in addition to three alternative platforms, with each array platform being deployed at three test sites. The important result of this study is the establishment of good comparability of data from different platforms and sites: quantitative measures across all one-colour array platforms had a median coefficient of variation (CV) of 5–15%, and a concordance rate for the qualitative calls ("Present," "Absent") of 80–95% between sample replicates, lists of differentially expressed genes overlapped by about 89% between the test sites using the same platform, dropping to 74% overlap across platforms. Consequently, the authors stress that standardization of data reporting, analysis tools and controls is important in this process. The two reference RNA samples A and B, which were used in the MAQC project, are commercially available and can be used to evaluate the laboratory proficiency during large gene expression studies.

The External RNA Control Consortium (ERCC) was established in 2003 with the goal to develop and establish a reference set of RNA which could be used to evaluate the technical performance of a gene expression measurement experiment by microarrays. About 100 well-characterized clones of RNA transcripts have been made available to the community and a proposed way of testing the controls is available online (http://www.NIST.gov) (34).

Pine and colleagues developed a test system in which the reference material consists of two mixes of different ratios of RNA from four different rat tissues (35). Hereby it is possible to measure whether ratios of various transcripts over time in the same laboratory are consistently different and whether the differences are equally well detected in different laboratories.

Still, to date there is no generally applicable consensus on quality parameters and best practices for experimental design, sample processing, data analysis, and interpretation.

4. Probe Annotation: The Real Target

In order to understand data from microarray experiments, researchers need to learn what target, e.g., gene, transcript, or exon, each probe or probe set interrogates on the microarray platform. This was a very significant challenge in the early days of microarray when individual core labs often made their own microarrays by spotting EST clones onto glass slides. EST clones are known to be prone to contamination and chimera issues on top of liquid handling and clone tracking errors introduced in the spotting process (36). Consequently, it is often hard to replicate the results from the earlier cDNA clone-based studies due to high noise levels and the uncertain probe identity under many situations.

The wide adoption of commercial microarray platforms in recent years not only greatly increases the consistency of microarray results but also significantly improves probe/probe set annotations, as teams of bioinformaticians are usually responsible for the selection of target and the design of probe/probe sets on various commercial array platforms. As a result, most end users of the commercial microarray platforms usually rely on the periodic annotation updates from the corresponding vendors, such as quarterly annotation updates from Affymetrix, Illumina, or Agilent, for linking microarray data points to specific genes, transcripts, or exons. Nonetheless, vendor-provided annotation updates can often be significantly improved to meet different data analysis requirements.

4.1. Probe/Probe Set Identity Issue

Unlike the Agilent custom eArray, the Affymetrix GeneChips and the Illumina BeadArrays usually are used for several years after the initial design. Due to the rapid increase in genome and transcriptome information in recent years, probes or probe sets thought to represent specific targets based on old information are often no longer correct at the time of experiment or data analysis. In fact, even for Agilent custom eArrays, reanalyzing data generated several years ago should use the most current genome/transcriptome data to improve analysis quality. In general, annotation updates from Illumina and Agilent are doing very well for providing correct probe identity to their corresponding platforms, as the one-probe-one-target design of these platforms makes re-annotation quite straightforward for data analysis.

However, re-annotation is more complicated for Affymetrix GeneChips designed for expression analysis, which is still the leading microarray platform as judged by number of data sets deposited in the NCBI GEO database and Medline records related to GeneChip assays. By design, Affymetrix uses multiple 25-mer probes to represent each target in order to obtain more reliable signal from short oligonucleotides. Although Affymetrix provides new annotations for each probe sets quarterly, either based on the sequence alignment results to the latest Entrez database using the representative sequences for probe sets or direct probe sequences, Affymetrix never updated their probe set definitions to remove problematic probes based on the latest genome/transcriptome information. Work from different groups as well as ours summarized several categories of problematic probes on some widely used GeneChips at that time, such as those from contaminating/chimeric EST sequences in the old UniGene database, probes with wrong identity, nonspecific probes or those overlapping with high-heterogeneity SNP sites (37–39).

Here we want to highlight some of the probe problems with the more recent Affymetrix Human Exon 1.0 Sense Transcript Array (HsExon Array). The release of HsExon Array in the second half of 2005 represented a major advancement in the ability to study transcriptome-wide splicing changes. According to Affymetrix, over five million sense probes on the HsExon Array are able to interrogate 1.4 million exons derived from different annotations (http://www.affymetrix.com/products_services/arrays/specific/exon_designstatistics.affx). Our analysis revealed two main problems with the HsExon Array probes:

1. About 6.9% of the probes on the HsExon Array can be aligned to multiple genomic locations. These nonspecific probes can influence the expression/splicing signal intensity of more than one transcript or exon, leading to mostly false positives and also false negatives under some situations. It can be estimated that about $1-(1-0.069)^4=25\%$ of the probe sets can potentially be influenced by the nonspecific probe problem, as the average size of a probe set representing an exon is 4 on the HsExon Array. Although not every nonspecific probe will alter analysis results, such nonspecific probes in exon probe set definitions will likely lower the analysis quality by reducing sensitivity and/or increase false positives.
2. Affymetrix' HsExon Array includes 271,710 probes with single genome hit but overlapping with over 57,000 known SNPs with minor allele frequency (MAF) above 0.1 in the central 15 bp region of the probe (Table 1). The real number of SNPs with MAF>0.1 is likely to be higher based on existing HapMap/ENCODE data (Table 1, columns 3 and 4) (40).

Based on the existing literature (41, 42), probes with central 15-bp mismatches are likely to generate very different signals in samples with different genotypes. The presence of such probes with relatively high MAF in a probe set may reduce the sensitivity of detecting real splicing changes due to increased noise level. It may also increase the false positive rate when the sample size is not very large due to the uneven allele distribution in different sample groups. For example, a MAF=0.2 allele has a 0.0029 probability of displaying ≥40%

Table 1
Number of SNPs represented by allele-specific probes on the Affymetrix HsExon Array

Minor allele frequency (MAF)	Number of known SNPs represented on the HsExon Array*			Estimated MAF percentage from HapMap/ENCODE (%)	Estimated SNP count on the HsExon Array based on HapMap/ENCODE data		
MAF > 0.10	57,532[#]	12,918[##]	4,472[###]	~41.5	93,007[#]	20,883[##]	7,229[###]
MAF > 0.20	40,606[#]	9,175[##]	3,212[###]	~26.5	59,390[#]	13,419[##]	4,698[###]
MAF > 0.30	26,353[#]	5,988[##]	2,100[###]	~16	35,858[#]	8,148[##]	2,857[###]

*Number of SNPs represented by a least one probe ([#]), two probes ([##]), and three probes ([###]) on the HsExon Array. Multiple probes often overlap due to inclusion of different gene/transcript/exon definitions

allele frequency difference in a 20 controls vs. 20 cases data set. Consequently, around 117 (40,606 known SNPs overlap with HsExon Array probes having MAF > 0.20 × 0.0029) SNPs will exhibit ≥40% frequency difference in these two groups by chance alone, creating ample opportunities for false exon-level signal differences given the average size of exon probe set is only four.

Since Affymetrix GeneChip data analysis depends on probe set definitions for summarizing signals from different probes in a probe set, just updating probe set annotations but keeping problematic probes in the original probe set definitions is a questionable approach. The most ideal solution is to remove the problematic probes from the original probe sets for more accurate estimation of the target signal levels during the data analysis process.

4.2. Target Selection Problem

A common target selection problem is the redundancy of probes/probe sets representing the same target, some of them can be flatly wrong. The main cause of this problem is the incompleteness of the UniGene database, which has been an important source for potentially novel gene-specific transcripts for many species, in particular for those without a completed genome. A large fraction of UniGene clusters are from the same genes due to insufficient sequencing depth from various EST sequencing projects. Table 2 is a summary of the chromosome and strand information for probe sets representing the FGFR2 receptor, which was identified as a critical gene involved in major depression by the Pritzker Neuropsychiatric Disorders Consortium (43). As one might expect, researchers were initially puzzled by strikingly different signals from the ten FGFR2 probe sets on the HG-U133A chip but a systematic mapping of probes to human genome revealed the origin of the inconsistency across different probe sets.

The Illumina whole genome arrays relied heavily on UniGene in their design thus they have the same redundancy issues as well as

Table 2
HG-133A FGFR2 probe sets have different strand and chromosome assignment

UniGene ID	Chip type	HG-U133 Acc	Probe set ID	Chromosome	Strand
Hs.533683	HG-U133A	AB030073	211400_at	10	–
Hs.533683	HG-U133A	AB030075	211398_at	10	+
Hs.533683	HG-U133A	AB030077	211399_at	10	+
Hs.533683	HG-U133A	AB030078	211401_s_at	10	–
Hs.533683	HG-U133A	M80634	203639_s_at	10	–
Hs.533683	HG-U133A	M87771	208228_s_at	10	–
Hs.533683	HG-U133A	NM_022969	203638_s_at	10	–
Hs.533683	HG-U133A	NM_022971	208225_at	1	+
Hs.533683	HG-U133A	NM_022975	208229_at	10	–
Hs.533683	HG-U133A	NM_022976	208234_x_at	10	–

other problems associated with the UniGene. In our re-annotation of the Illumina Mouse Whole Genome-6 v1_1 bead array for internal research projects, we noticed about 40% of the probes do not match the current mouse genome and transcript sequences even if when we allow one base mismatch for each probe. Around a third of the valid target genes are represented by more than one probe on the Illumina whole genome bead arrays, which are supposed to use one probe sequence to represent one gene. A recent paper on an Illumina BeadArray re-annotation pipeline also described similar problems (44).

The HsExon Array mentioned previously has another type of target selection issue: it includes many different transcript/exon annotations with different level of reliability. Affymetrix groups probe sets into three categories (core, extended and full), from the most conservative to highly exploratory, for maximizing the chance of discovering novel splicing and expression patterns (45). However, while the inclusion of all possible exons derived from different annotation methods is a good strategy for dealing with lack of alternative splicing information, it leads to difficulties in the interpretation of analysis results since most of the sources used by Affymetrix do not provide straightforward Web-based visualization for the corresponding exons in the context of other genome structures. In fact, Affymetrix' gene/transcript/exon definitions are not widely adopted by the research community and none of the three Affymetrix probe set categories is 100% consistent with popular genome annotation databases such as ENSEMBL or Entrez Gene. There is high level of overlapping among different exon definitions used by HsExon Array, too. It will be desirable to use individual high-quality genome annotation databases in the data analysis stage for more straightforward interpretation. Such a re-annotation will also help the multiple testing problem significantly: the latest ENSEMBL Exon-based annotation of the HsExon Array has 304,497 exon probe sets, which is less than 1/4 of the 1.4 million exon probe sets from different sources in Affymetrix' original design.

Conceivably, besides removing problematic probes from the original probe sets, individual probes hitting the same target (gene, transcript, or exon) should be grouped together in new probe sets, which will likely reduce the noise level due to larger probe set sizes as well as increase the statistical power of data analysis by reducing the number of probes/targets.

4.3. Re-annotation of Commercial Microarrays

Given the probe/probe set and target selection problems on widely used microarray platforms, it is necessary to re-annotate the original probe/probe sets using the most updated genome and transcriptome information. We generated the so-called "custom CDF" by setting up a GeneChip re-annotation pipeline that will realign GeneChip probes to the latest genome and transcriptome sequences, remove problematic probes using a series of filters, combine good probes representing the same target into probe sets based on widely used gene, transcript, and exon definitions.

We decided to annotate the same GeneChip based on multiple gene, transcript, and exon definitions since there are significant inconsistencies across major databases such as UniGene, Entrez Gene, Refseq, ENSEMBL, and VEGA. Each of the definitions has its own pros and cons, and users can select the probe set definitions most suitable for their analysis needs. For example, if a researcher is interested in obtaining the best coverage of the human transcriptome, our recommendation in the early days is the UniGene-based probe sets since UniGene-based annotations have the largest probe set count. Now we usually recommend Entrez Gene for human and mouse due to its better stability and wide usage, in addition to the fact that the number of Entrez-based probe sets is on par with the UniGene-based probe sets in the last couple of years. Refseq-based CDFs are the top choice if researchers are interested in having the best stability of their analysis results in the long run since individual reference sequences hardly change across time while new reference sequences are added. Naturally, exon-based CDFs are needed if a user is interested in exploring potential alternative splicing events.

We call such updated annotation targeted for different user requirements the "Custom CDFs" and they can be downloaded automatically during in R sessions using custom CDF. Individual custom CDF packages can also be downloaded through the link at the BioConductor Web site (http://www.bioconductor.org/download) as well as our own Web site (http://brainarray.mbni.med.umich.edu/Brainarray/Database/CustomCDF/CDF_download.asp).

The details of our re-annotation pipeline and how to use custom CDF are described in Dai et al. (37).

Besides custom CDF packages, we provide a number of tools to facilitate GeneChip data reanalysis using custom CDFs including (1) The Web-Based GeneChip Analysis System (WGAS): it enables the use of custom CDFs (as well as Affymetrix' original CDFs) for the reanalysis of public domain CEL files deposited in the GEO database with a number of popular GeneChip analysis algorithms such as RMA, dChip, and MAS5 (46). The straightforward GUI interface provided by WGAS allows researchers to setup basic reanalysis tasks without the need to type any R code although advanced users can modify the automatically generated R-code freely to tailor the analysis to meet their special requirements. (2) A Web function for users to query the exact probes in each probe set and its genomic location, (3) downloadable R packages for uploading CEL files to WGAS and for performing custom probe filtering and cross GeneChip analysis (46).

The custom CDFs are now in its 12th version and our pipeline covers GeneChip from 16 species. Many users offered helpful suggestions for improving the custom CDF annotation. In addition, we rely on users' help for the annotation of nonmammalian species as well as the GeneChip ST arrays that we do not have direct experience. Our re-annotation work has been cited by over 200 papers. Most importantly, two papers published soon after our 2005 Nucleic Acid Research paper described systematic comparative analysis of our custom CDF and the original Affymetrix probe set definition (47, 48). Lu and Zhang discovered that "The false-positive discoveries with the AFFY definition are double of those with the DUG definition (AFFY is Affymetrix' original probe set definition and DUG refers to our UniGene-based custom CDF)" and they concluded that "… adjusting the criteria for analyzing the genes based on the old probe-set definition is not a solution for compensating for the effect of less-accurate probe mapping" (46). Sandberg and Larsson performed a systematic comparison of different custom CDFs we created vs. Affymetrix CDF and suggested that "Updated probe set definitions do not only offer expression levels that are more accurately associated to genes and transcripts but also shows improvements in the estimated transcript expression levels. These results give further support for a widespread use of updated probe set definitions for analysis and meta-analysis of microarray data" (48).

4.4. Additional Transcription and Genome Structure Information Through Target-Specific Annotations

Improving microarray annotation will not only enable more accurate interpretation of results by reducing the flaws in the original array designs, but it also provides the possibility of extracting additional information, such as alternative splicing or genomic structure alteration information from existing microarray data. The inclusion of many probes across the transcriptome at least in theory

Table 3
Representation of multiple exons in non-exon array GeneChips

GeneChip	Total ENSG	Total ENSE	Two ENSEs	Three ENSEs	≥4 ENSEs
HG-U133 Plus2	18,394	46,321	4,584	2,347	3,685
MG 430 2.0	15,041	27,827	4,152	1,635	1,310
MG 230 2.0	9,023	13,695	2,123	619	378
HsGene 1.0 ST v1	22,197	91,170	3,084	2,835	9,964
MmGene 1.0 ST v1	20,963	84,144	2,539	1,684	8,634
RnGene 1.0 ST v1	21,130	74,196	2,781	1,847	7,783

Table 4
Database of genomic variation locus represented on GeneChips

GeneChip	Unique variation loci	Variation ID-based probe sets	Probes used
HGU133A	1,196	3,236	62,073
HGU133B	1,271	3,359	61,921
MG 230 2.0	1,770	4,945	152,963
HGU133Plus2	4,325	12,043	1,416,513
HuEx10stv2	2,031	5,751	226,012

provides the possibility of detecting of some alternative splicing events as well as copy number variations in genome. Table 3 is a summary of multiple exons that Affymetrix GeneChips can detect despite the fact that the original arrays were not designed for such purposes.

Table 4 is the count of CNVs that GeneChips may detect based on CNV information in the Database of Genomic Variations (DGV) hg18_v4 (49). We would like to point out that the total number of unique variation loci was 5,083 in that release; thus existing human GeneChips already cover a fairly significant portion of unique variation loci. It is also worth mentioning that GeneChip designed for SNP genotyping was used for the alternative purpose of detecting allele-specific expression, and it was actually the first genome-wide assessment of allele-specific expression at its time (50).

Although such alternative use of GeneChips may detect only a relative small portion of all possible alternative splicing events or CNVs, the fact that no additional experiment is needed for finding novel splicing and CNV candidates related to the targeted pathophysiological process is very attractive. All users need to do is to reanalyze existing data with new annotations designed for a specific purposes.

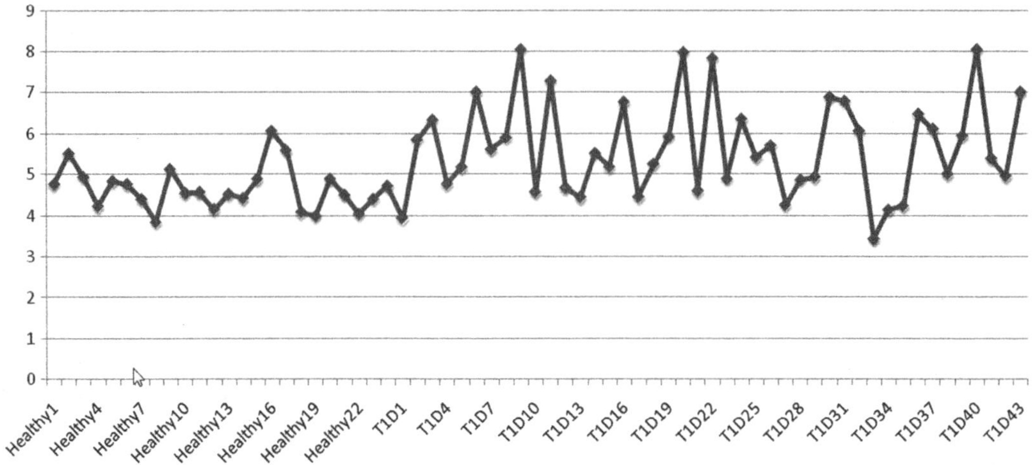

Fig. 2. Signal of DGV Variation_32230 in healthy and T1D blood samples.

Figure 2 is an example of CNV analysis results using a public domain HG-U133 data set (GSE9006) intended for detecting differential gene expression in peripheral blood mononuclear cells between healthy and type 1 diabetes (T1D) children (51). It can be seen that (1) DGV Variation_32230 signals exhibit a bi-allelic-like pattern, with lowest signals around 4, middle level signals around 6 and highest level signals around 8. (2) High signals (above 7) only show up in T1D samples but not in healthy samples. Such a signal distribution suggests possible higher copy number (2 copies or possibly more) of this region in some T1D subjects. It will likely be worthwhile to follow up this array-based discovery with genomic and CYP4F3 mRNA PCR to see if this indeed involves genomic CNV or it is simply due to higher expression level of CYP4F3 expression in some T1D samples.

Conceivably, the Illumina BeadArrays and the Agilent eArrays can also be annotated for the extraction of splicing and CNV-related information. Standard GeneChip analysis algorithms such as RMA and MAS5 can be used in a straightforward manner for summarizing signals from multiple probes if a CNV region or transcript is represented by more than one probe.

5. Data Standardization

A typical gene expression experiment not only results in raw data but is associated with a large number of diverse data. The data structure, which contains raw data and data-like design, sample information, protocols, medical information, is often referred to as "metadata," and needs to be recorded, reported, and stored. The ultimate goal is the integration of this knowledge in the description and modeling of biological systems (52). Quality of data reporting is an essential requirement for ensuring reproducibility of results. Public databases have been designed for the purpose of data storage such as Gene Expression Omnibus (http://www.ncbi.nlm.nih.gov/geo) or ArrayExpress (http://www.ebi.ac.uk/microarray-as/ae). Also, many scientific journals require now the deposition of data before considering publication of the results. However, the format, level, and degree of data caption is not standardized and leaves room for improvement to an extent that results are not reproducible by external groups (26). Sharing data per se does not seem to be useful. Efficient data analysis requires understanding the structure of the underlying metadata which is why simply depositing raw data is not sufficient. Rather, metadata need to be reported in an exhaustive and standardized way in a public database.

Over the years the importance of standardization of data sharing infrastructure as well of the development of standardized tools for analysis has been stressed (26, 53–55). The development of such an infrastructure should aim at providing guidelines with respect to information content, syntax (format), and semantics (e.g., ontology, metadata). Several approaches already exist. The concept of Minimum Information (MI) content is still under discussion. MI content should be provided when describing experiments and reporting data. However, there is still no conclusion on which information is "minimal" and how extensive the content needs to be. The issue of file formats is a balance of accessibility/usability and computer readability. Gentleman has proposed that authors should publish their data as a "compendium". A compendium should include every piece of information needed to reproduce the analysis, such as text, codes, and data (56). In contrast to a static published manuscript, the "dynamic document" would incorporate all of the author's work and allow different levels of detail to be displayed in derived documents, the scripts are easy to rerun, provide explicit computational details, enable the generation of plots, and allow the document to be treated as data (56). Additionally, the compendium provides check points enabling the reader to analyze the reproducibility of the published data. In conjunction, the development of the Bioconductor project was a major step towards a harmonization of analytical software (www.bioconductor.org). The third aspect, ontology, refers to a formal specification of terms in a particular subject area and the relations among them. Gene Ontology (GO; (57)) is a controlled vocabulary used to describe the biology of a gene product in any organism. Each gene is assigned three independent terms describing: (1) the molecular function of the gene product; (2) the biological process in which the gene product participates; and (3) the cellular component where the gene product can be found. To be successful, one must create an ontology that is well defined yet reasonable in size but that does not overly limit our ability to describe an experiment. Recently, the GO consortium introduced a Reference Genome Project with the goal to increase annotation consistency across genome databases and improve the logical structure and biological content of the "traditional" GO (58). It will be interesting to follow these developments.

Let us have a look at some of the organizations which attempt to address data standardization.

The Microarray and Gene Expression Data Society (MGED) was one of the first initiatives to develop guidelines for description and storage of microarray data. The creation of the Minimal Information About a Microarray Experiment (MIAME) checklist and standards for microarray data annotation (59) was only one success of MGED. MIAME is the most advanced standard initiative and is considered the benchmark for the development of new standards. Another success is the establishment of the MicroArray Gene Expression Object Model (MAGE-OM) and its XML-based implementation (MAGE-ML (60)). The MAGE-OM describes information about microarray-based experiments. The complexity of MAGE-ML made it difficult for biologists to use it and a simpler spreadsheet format, MAGE-TAB, is now recommended for representing microarray data and metadata (61). Another example of the fruitful work of MGED is the introduction of the MGED Ontology, as an annotation resource for microarray data (62). As a consequence of their efforts, MIAME standards are now widely accepted by the microarray community. In support of MIAME and the associated publication requirements, MGED worked with the three primary public biological data repositories and their gene-expression databases (63, 64): ArrayExpress (65), Gene Expression Omnibus (66) (GEO), and CIBEX (67).

Standardization efforts for data for clinical applications are also called upon by governmental regulatory agencies, like the EPA and the FDA, with particular emphasis put on best practices for sampling and reporting of clinical trials data (68). With the major contribution of the FDA, Clinical Data Interchange Standards Consortium (CDISC) has become an important organization for data standardization within the pharmaceutical and clinical research communities (www.cdisc.org). Under the umbrella of CDISC, standards for all aspects of drug-development have been developed, including trial design, data collection, storage, sharing, and reporting (69).

The EPA document "Interim Guidance for Microarray-Based Assays: Data submission, Quality, Analysis, Management, and Training Considerations" (70) is an important step towards harmonization of microarray data generation and interpretation. The authors purposefully do not recommend any specific methods or quality control measures, but rather enforce the importance of experimental design and performance measures like positive and/or negative controls, and the inclusion of calibration curves with external RNA standards, supported by a complete-as-possible documentation.

6. Conclusion

The generation of quality data, which are reliable, comparable, and testable, can be achieved by following good experimental practice and by providing details on the experiment and specimen in a most accurate and thorough way, meeting terminological criteria of annotation standards, such as MIAME. The number of standards calls for a broad community discussion on "standardizing the standards," (52) and on a common agreement with financial funding to further develop a commonly applicable standard.

References

1. Schena M, Shalon D, Davis RW, Brown PO (1995) Quantitative monitoring of gene expression patterns with a complementary DNA microarray. Science 270:467–470
2. Lockhart DJ, Dong H, Byrne MC, Follettie MT, Gallo MV, Chee MS, Mittmann M, Wang C, Kobayashi M, Horton H, Brown EL (1996) Expression monitoring by hybridization to high-density oligonucleotide arrays. Nat Biotechnol 14:1675–1680
3. Brown PO, Botstein D (1999) Exploring the new world of the genome with DNA microarrays. Nat Genet 21:33–37
4. Rogers S, Cambrosio A (2007) Making a new technology work: the standardization and regulation of microarrays. Yale J Biol Med 80:165–178
5. The Tumor Analysis Best practices Working Group (2004) Expression profiling-best practices for data generation and interpretation in clinical trials. Nat Rev 5:229–237
6. http://www.fda.gov/downloads/Drugs/GuidanceComplianceRegulatoryInformation/Guidances/UCM079855.pdf
7. Shi L, Reid LH, Jones WD, Shippy R, Warrington JA, Baker SC, Collins PJ, de Longueville F, Kawasaki ES, Lee KY, Luo Y, Sun YA, Willey JC, Setterquist RA, Fischer GM, Tong W, Dragan YP, Dix DJ et al (2006) The MicroArray Quality Control (MAQC) project shows inter- and intraplatform reproducibility of gene expression measurements. Nat Biotechnol 24:1151–1161
8. Clarke JD, Zhu T (2006) Microarray analysis of the transcriptome as a stepping stone towards understanding biological systems: practical considerations and perspectives. Plant J 45:630–650
9. Dix DJ, Gallagher K, Benson WH, Groskinsky BL, McClintock T, Dearfield KL, Farland WH (2006) A framework for the use of genomics data at the EPA. Nat Biotechnol 24:1108–1111
10. Grass P (2009) Experimental design, pp 19–31. In: Scherer A (ed) Batch effects and noise in microarray experiments. Wiley, West Sussex, ISBN:978-0-470-74138-2
11. Sica GT (2006) Bias in research studies. Radiology 238:780–789
12. Rudic RD, McNamara P, Reilly D, Grosser T, Curtis AM, Price TS, Panda S, Hogenesch JB, FitzGerald GA (2005) Bioinformatic analysis of circadian gene oscillation in mouse aorta. Circulation 112:2716–2724
13. Coombes KR, Highsmith WE, Krogmann TA, Baggerly KA, Stivers DN, Abruzzo LV (2002) Identifying and quantifying sources of variation in microarray data using high-density cDNA membrane arrays. J Comp Biol 9:655–669
14. Li X, Gu WMS, Balink D (2002) DNA microarrays: their use and misuse. Microcirculation 9:13–22
15. Zakharkin SO, Kim K, Mehta T, Chen L, Barnes S, Scheirer KE, Parrish RS, Allison DB, Page GP (2005) Sources of variation in

Affymetrix microarray experiments. BMC Bioinform 6:214

16. Auer H, Lyianarachchi S, Newsom D, Klisovic MI, Marcucci G, Kornacker K (2003) Chipping away at the chip bias: RNA degradation in microarray analysis. Nat Genet 35:292–293
17. Dumur CI, Nasim S, Best AM, Archer KJ, Ladd AC, Mas VR, Wilkinson DS, Garrett CT, Ferreira-Gonzalez A (2006) Evaluation of quality-control criteria for microarray gene expression analysis. Clin Chem 50:1994–2002
18. Schroeder A, Mueller O, Stocker S, Salowsky R, Leiber M, Gassmann M, Lightfoot S, Menzel W, Granzow M, Ragg T (2006) The RIN: an RNA integrity number for assigning integrity values to RNA measurements. BMC Mol Biol 7:3
19. Ioannidis JP (2005) Microarrays and molecular research: noise discovery? Lancet 365:454–455
20. Frantz S (2005) An array of problems. Nat Rev Drug Discov 4:362–363
21. Strauss E (2006) Arrays of hope. Cell 127:657–659
22. Ying L, Sarwal M (2008) In praise of arrays. Pediatr Nephrol 24:1643–1659
23. Marshall E (2004) Getting the noise out of gene arrays. Science 306:630–631
24. Michiels S, Koscielny S, Hill C (2005) Prediction of cancer outcome with microarrays: a multiple random validation strategy. Lancet 365:488–492
25. Ein-Dor L, Zuk O, Domany E (2006) Thousands of samples are needed to generate a robust gene list for predicting outcome in cancer. Proc Natl Acad Sci USA 103:5923–5928
26. Ioannidis JP, Allison DB, Ball CA, Coulibaly I, Cui X, Culhane AC, Falchi M, Furlanello C, Game L, Jurman G, Mangion J, Mehta T, Nitzberg M, Page GP, Petretto E, van Noort V (2009) Repeatability of published microarray gene expression analyses. Nat Genet 41:149–155
27. Dobbin KK, Beer DG, Meverson M, Yeatman TJ, Gerald WL, Jacobson JW, Conley B, Buetow KH, Heiskanen M, Simon RM, Minna JD, Girard L, Misek DE, Taylor JM, Hanash S, Naoki K, Hayes DN, Ladd-Acosta C, Enkemann SA, Viale A, Giordano TJ (2005) Interlaboratory comparability study of cancer gene expression analysis using oligonucleotide microarrays. Clin Cancer Res 11:565–572
28. Larkin JE, Frank BC, Gavras H, Sultana R, Quackenbush J (2005) Independence and reproducibility across microarray platforms. Nat Methods 2:337–344
29. Irizarry RA, Warren D, Spencer F, Kim IF, Biswal S, Frank BC, Gabrielson E, Garcia JG, Geoghegan J, Germino G, Griffin C, Hilmer SC, Hoffman E, Jedlicka AE, Kawasaki E, Martínez-Murillo F, Morsberger L, Lee H, Petersen D, Quackenbush J, Scott A, Wilson M, Yang Y, Ye SQ, Yu W (2005) Multiple-laboratory comparison of micrarray platforms. Nat Methods 2:345–350
30. Chudin E, Walker R, Kosaka A, Wu SX, Rabert D, Chang TK, Kreder DE (2002) Assessment of the relationship between signal intensities and transcript concentration for Affymetrix GeneChip arrays. Genome Biol 3:RESEARCH0005
31. Moreau Y, Aerts S, De Moor B, De Strooper B, Dabrowski M (2003) Comparison and meta-analysis of microarray data: from the bench to the computer desk. Trends Genet 19:570–577
32. Kim H, Zhao B, Snesrud EC, Haas BJ, Town CD, Quackenbush J (2002) Use of RNA and genomics DNA references for inferred comparisons in DNA microarray analyses. Biotechiques 33:924–930
33. Miklos GL, Maleszka R (2004) Microarray reality checks in the context of a complex disease. Nat Biotechnol 22:615–621
34. The External RNA Controls Consortium (2005) The external RNA controls consortium: a progress report. Nat Methods 2:731–734
35. Pine PS, Boedigheimer M, Rosenzweig BA, Turpaz Y, He YD, delestarr G, Ganter B, Jarnagin K, Jones WD, Reid LH, Thompson KL (2008) Use of disganostic accuracy as a metric for evaluating laboratory proficiency with microarray assays using mixed-tissue RNA reference samples. Pharmacogenomics 9: 1753–1763
36. Halgren RG, Fielden MR, Fong CJ, Zacharewski TR (2001) Assessment of clone identity and sequence fidelity for 1189 IMAGE cDNA clones. Nucleic Acids Res 29:582–588
37. Dai M, Wang P, Boyd AD, Kostov G, Athey B, Jones EG, Bunney WE, Myers RM, Speed TP, Akil H, Watson SJ, Meng F (2005) Evolving gene/transcript definitions significantly alter the interpretation of GeneChip data. Nucleic Acids Res 33:e175
38. Gautier L, Moller M, Friis-Hansen L, Knudsen S (2004) Alternative mapping of probes to genes for Affymetrix chips. BMC Bioinform 5:111
39. Harbig J, Sprinkle R, Enkemann SA (2005) A sequence-based identification of the genes detected by probesets on the Affymetrix U133 plus 2.0 array. Nucleic Acids Res 33:e31
40. Altshuler D, Brooks LD, Chakravarti A, Collins FS, Daly MJ, Donnelly P (2005) A haplotype map of the human genome. Nature 437: 1299–1320

41. Lee I, Dombkowski AA, Athey BD (2004) Guidelines for incorporating non-perfectly matched oligonucleotides into target-specific hybridization probes for a DNA microarray. Nucleic Acids Res 32:681–690
42. Mei R, Hubbell E, Bekiranov S, Mittmann M, Christians FC, Shen MM, Lu G, Fang J, Liu WM, Ryder T, Kaplan P, Kulp D, Webster TA (2003) Probe selection for high-density oligonucleotide arrays. Proc Natl Acad Sci USA 100:11237–11242
43. Evans SJ, Choudary PV, Neal CR, Li JZ, Vawter MP, Tomita H, Lopez JF, Thompson RC, Meng F, Stead JD, Walsh DM, Myers RM, Bunney WE, Watson SJ, Jones EG, Akil H (2004) Dysregulation of the fibroblast growth factor system in major depression. Proc Natl Acad Sci USA 101:15506–15511
44. Barbosa-Morais NL, Dunning MJ, Samarajiwa SA, Darot JF, Ritchie ME, Lynch AG, Tavare S (2010) A re-annotation pipeline for Illumina BeadArrays: improving the interpretation of gene expression data. Nucleic Acids Res 38(3):e17
45. Affymetrix (2005) (a) Exon Array Computational Tool Software User's Guide, (b) Whole Transcript (WT) Sense Target Labeling Assay Manual, (c) Alternative Transcript Analysis Methods for Exon Arrays v1.1, (d) Exon Array Background Correction v1.0, (e) Exon Probeset Annotations and Transcript Cluster Groupings v1.0, (f) Gene Signal Estimates from Exon Arrays v1.0, (g) Quality Assessment of Exon Arrays v1.0', http://www.affymetrix.com/products/arrays/specific/exon.affx, Human Exon 1.0 ST Array Manuals and White Papers
46. Dai M, Wang P, Jakupovic E, Watson SJ, Meng F (2007) Web-based GeneChip analysis system for large-scale collaborative projects. Bioinformatics 23:2185–2187
47. Lu X, Zhang X (2006) The effect of GeneChip gene definitions on the microarray study of cancers. Bioessays 28:739–746
48. Sandberg R, Larsson O (2007) Improved precision and accuracy for microarrays using updated probe set definitions. BMC Bioinform 8:48
49. Iafrate AJ, Feuk L, Rivera MN, Listewnik ML, Donahoe PK, Qi Y, Scherer SW, Lee C (2004) Detection of large-scale variation in the human genome. Nat Genet 36:949–951
50. Lo HS, Wang Z, Hu Y, Yang HH, Gere S, Buetow KH, Lee MP (2003) Allelic variation in gene expression is common in the human genome. Genome Res 13:1855–1862
51. Kaizer EC, Glaser CL, Chaussabel D, Banchereau J, Pascual V, White PC (2007) Gene expression in peripheral blood mononuclear cells from children with diabetes. J Clin Endocrinol Metab 92:3705–3711
52. Ideker T, Galitski T, Hood L (2001) A new approach to decoding life: systems biology. Annu Rev Genomics Hum Genet 2:343–372
53. Quackenbush J (2006) Standardizing the standards. Mol Syst Biol 2:2006.0010
54. Williams-Devane CR, Wolf MA, Richard AM (2009) Toward a public toxicogenomics capability for supporting predictive toxicology: survey of current resources and chemical indexing of experiments in GEO and ArrayExpress. Toxicol Sci 109:358–371
55. CASIMIR Consortium (2009) Post-publication sharing of data and tools. Nature 461:171–173
56. Gentleman R, Lang DT (2004) Statistical analyses and reproducible research. www.bepress.com/bioconductor/paper2
57. Ashburner M, Ball C, Blake J, Botstein D, Butler H, Cherry JM, Davis AP, Dolinski K, Dwight SS, Eppig JT, Harris MA, Hill DP, Issel-Tarver L, Kasarskis A, Lewis S, Matese JC, Richardson JE, Ringwald M, Rubin GM, Sherlock G (2000) Gene ontology: tool for the unification of biology. The Gene Ontology Consortium. Nat Genet 25:25–32
58. Gaudet P, Chisholm R, Berardini T, Dimmer E, Engel SR, Fey P, Hill DP, Howe D, Hu JC, Huntley R, Khodiyar VK, Kishore R, Li D, Lovering RC, McCarthy F, Ni L, Petri V, Siegele DA, Tweedie S, Van Auken K, Wood V, Basu S, Carbon S, Dolan M, Mungall CJ, Dolinski K, Thomas P, Ashburner M, Blake JA, Cherry JM, Lewis SE, Balakrishnan R, Christie KR, Costanzo MC, Deegan J, Diehl AD, Drabkin H, Fisk DG, Harris M, Hirschman JE, Hong EL, Ireland A, Lomax J, Nash RS, Park J, Sitnikov D, Skrzypek MS, Apweiler R, Bult C, Eppig J, Jacob H, Parkhill J, Rhee S, Ringwald M, Sternberg P, Talmud P, Twigger S, Westerfield M (2009) The Gene Ontology's Reference Genome Project: a unified framework for functional annotation across species. PLoS Comput Biol 5:e1000431
59. Brazma A, Hingamp P, Quackenbush J, Sherlock G, Spellman P, Stoeckert C, Aach J, Ansorge W, Ball CA, Causton HC, Gaasterland T, Glenisson P, Holstege FC, Kim IF, Markowitz V, Matese JC, Parkinson H, Robinson A, Sarkans U, Schulze-Kremer S, Stewart J, Taylor R, Vilo J, Vingron M (2001) Minimum information about a microarray experiment (MIAME)-toward standards for microarray data. Nat Genet 29:365–371
60. Spellman PT, Miller M, Stewart J, Troup C, Sarkans U, Chervitz S, Bernhart D, Sherlock G,

Ball C, Lepage M, Swiatek M, Marks WL, Goncalves J, Markel S, Iordan D, Shojatalab M, Pizarro A, White J, Hubley R, Deutsch E, Senger M, Aronow BJ, Robinson A, Bassett D, Stoeckert CJ Jr, Brazma A (2002) Design and implementation of microarray gene expression markup language (MAGE-ML). Genome Biol 3:RESEARCH0046

61. Rayner TF, Rocca-Serra P, Spellman PT, Causton HC, Farne A, Holloway E, Irizarry RA, Liu J, Maier DS, Miller M, Petersen K, Quackenbush J, Sherlock G, Stoeckert CJ Jr, White J, Whetzel PL, Wymore F, Parkinson H, Sarkans U, Ball CA, Brazma A (2006) A simple spreadsheet-based, MIAME-supportive format for microarray data: MAGE-TAB. BMC Bioinform 7:489

62. Whetzel PL, Parkinson H, Causton HC, Fan L, Fostel J, Fragoso G, Game L, Heiskanen M, Morrison N, Rocca-Serra P, Sansone SA, Taylor C, White J, Stoeckert CJ Jr (2006) The MGED Ontology: a resource for semantics-based description of microarray experiments. Bioinformatics 22:866–873

63. Ball CA, Brazma A, Causton H, Chervitz S, Edgar R, Hingamp P, Matese JC, Parkinson H, Quackenbush J, Ringwald M, Sansone SA, Sherlock G, Spellman P, Stoeckert C, Tateno Y, Taylor R, White J, Winegarden N (2004) Standards for microarray data: an open letter. Environ Health Perspect 112:A666–A667

64. Ball CA, Brazma A, Causton H, Chervitz S, Edgar R, Hingamp P, Matese JC, Parkinson H, Quackenbush J, Ringwald M, Sansone SA, Sherlock G, Spellman P, Stoeckert C, Tateno Y, Taylor R, White J, Winegarden N (2004) Submission of microarray data to public repositories. PLoS Biol 2:E317

65. Parkinson H, Kapushesky M, Shojatalab M, Abeygunawardena N, Coulson R, Farne A, Holloway E, Kolesnykov N, Lilja P, Lukk M, Mani R, Rayner T, Sharma A, William E, Sarkans U, Brazma A (2007) ArrayExpress—a public database of microarray experiments and gene expression profiles. Nucleic Acids Res 35:D747–D750

66. Edgar R, Domrachev M, Lash AE (2002) Gene Expression Omnibus: NCBI gene expression and hybridization array data repository. Nucleic Acids Res 30:207–210

67. Ikeo K, Ishi-i J, Tamura T, Gojobori T, Tateno Y (2003) CIBEX: center for information biology gene expression database. C R Biol 326:1079–1082

68. Frueh F (2006) Impact of microarray data quality on genomic data submissions to the FDA. Nat Biotechnol 24:1105–1107

69. Souza T, Kush R, Evans JP (2007) Global clinical data interchange standards are here! Drug Discov Today 12:174–181

70. U.S. Environmental Protection Agency DRAFT: Interim Guidance for Microarray-Based Assays: Data Submission, Quality, Analysis, Management, and Training Considerations (2007). http://www.epa.gov/osa/spc/pdfs/epa_interim_guidance_for_microarray-based_assays-external-review_draft.pdf

Chapter 11

Aggregation Effect in Microarray Data Analysis

Linlin Chen, Anthony Almudevar, and Lev Klebanov

Abstract

Inferring gene regulatory networks from microarray data has become a popular activity in recent years, resulting in an ever-increasing volume of publications. There are many pitfalls in network analysis that remain either unnoticed or scantily understood. A critical discussion of such pitfalls is long overdue. Here we discuss one feature of microarray data the investigators need to be aware of when embarking on a study of putative associations between elements of networks and pathways.

Key words: Microarray data, Networks, Pathways, Genetic pathways

1. Aggregated Expression Intensities

In a paper (1) Chu et al. pointed out the important fact that the measurements of mRNA abundance produced by microarray technology represent aggregated expression signals and, as such, may not adequately reflect the molecular events occurring within individual cells. To illustrate this conjecture, the authors proceeded from the observation that each gene expression measurement produced by a microarray is of the sum of the expression levels over many cells. Let ν be the number of cells contributing to the observed expression signal U (see Remark 1 below) and denote by X_i the expression level of a given gene in the ith cell. The notation Y_i is used for the second gene in a given pair of genes.

A simplistic model of the observed expression signals in this pair is given by

$$U = \sum_{i=1}^{\nu} X_i, \quad V = \sum_{i=1}^{\nu} Y_i, \qquad (1)$$

where X_i and Y_i are two sequences of independent and identically distributed (i.i.d.) random variables (r.v.s), while X_i and Y_i in each pair (X_i, Y_i) may be dependent with joint distribution function $F(x, y)$.

Andrei Y. Yakovlev et al. (eds.), *Statistical Methods for Microarray Data Analysis: Methods and Protocols*, Methods in Molecular Biology, vol. 972, DOI 10.1007/978-1-60327-337-4_11, © Springer Science+Business Media New York 2013

Limiting themselves to the case where ν is nonrandom, Chu et al. showed that, except for some very special and biologically irrelevant cases, the Markov factorization admitted by the expression levels within individual cells does not survive the summation (aggregation) in formula (1). The importance of this observation cannot be emphasized enough. However, as apparent from the relevant literature, it went entirely unnoticed.

In their concluding remarks, Chu et al. note that the mean vector and covariance matrix remain "invariant under aggregation up to a simple linear transformation." The same is obviously true for the correlation matrix.

Unfortunately, this is deflated when considering the case of random ν. Indeed, let each X_i have the same distribution as X, while each Y_i is distributed as Y. Then the following formula holds for the correlation coefficient $\rho(U,V)$ between U and V:

$$\rho(U,V) = \frac{\mu_\nu \mathrm{Cov}(X,Y) + \sigma_\nu^2 \mu_x \mu_y}{\sqrt{\mu_\nu \sigma_x^2 + \sigma_\nu^2 \mu_x^2} + \sqrt{\mu_\nu \sigma_y^2 + \sigma_\nu^2 \mu_y^2}}, \tag{2}$$

where $\mu_\nu = \mathrm{E}(\nu)$, $\mu_x = \mathrm{E}(X)$, $\mu_y = \mathrm{E}(Y)$,

$\sigma_\nu^2 = \mathrm{Var}(\nu)$, $\sigma_x^2 = \mathrm{Var}(X)$, $\sigma_y^2 = \mathrm{Var}(Y)$

and $\mathrm{Cov}(X,Y)$ is the covariance between X and Y.

Formula (2) can be represented as

$$\rho(U,V) = \frac{\rho(X,Y)}{\sqrt{1+a^2\tau}} \frac{1}{\sqrt{1+b^2\tau}} + \frac{\tau ab}{\sqrt{1+a^2\tau}\sqrt{1+b^2\tau}}, \tag{3}$$

where $\tau = \sigma_\nu^2 / \mu_\nu$, $a = \mu_x / \sigma_x$, $b = \mu_y / \sigma_y$, and $r = \rho(X,Y)$ ist the coefficient of correlation between X and Y. Therefore, $\rho(U,V) = \rho(X,Y)$ if and only if $\sigma_\nu = 0$.

Remark 1. If the hybridization reaction reaches equilibrium, an assumption widely adopted in the physical chemistry of microarrays, the random variable (r.v.) ν can be interpreted as the total number, N, of cells from which the total RNA is extracted. In the practical use of microarray technology, however, the reaction is typically stopped before equilibrium has been reached. In the latter case, the r.v. ν represents the number of cells that collectively yield the ultimate number of bound target–probe duplexes. Therefore, the random parameter ν is unobservable and should be thought of as a virtual number of cells associated with each batch of target RNA produced by them. This notion provides a constructive way of bridging the processes of gene expression at the genomic and tissue levels, which is the main thrust of our discussion. The conventional protocol of a microarray experiment implies that it is the total amount of RNA that is controlled (kept constant) across the arrays (subjects) rather than the number of cells ending up on each array. Therefore, the random fluctuations of ν cannot be controlled directly. Even if a tight control of N could be provided in experiments, it is unclear whether this would have had a diminishing effect on the variance of ν.

An upper bound for the deviation between $\rho(U,V)$ and $\rho(X,\Upsilon)$ is given by

$$|\rho(U,V)-\rho(X,\Upsilon)|\leq\frac{1}{2}\tau\left((a+b)^2+a^2b^2\tau\right), \tag{4}$$

This result follows from formula (3). Recall that the equality $\rho(U,V)=\rho(X,\Upsilon)$ holds when $\tau=0$. Considering $R=\rho(U,V)$ as a function of τ, one can verify that $R(\tau)$ either increases monotonically or attains a minimum before starting to increase with increasing τ. In both cases, $R\to 1$ when $\tau\to\infty$. The function $R(\tau)$ is smooth at $\tau=0$, but its initial slope may be quite high as our sample computations show. An additional quantitative insight into the potential impact of this unobservable variation on the correlation structure of microarray data is possible.

Remark 2. Like Chu et al., we consider the usual notion of correlation as a characteristic of the joint distribution of two r.v.s. Whenever the r.v.s are directly observable, a consistent estimator of the population correlation coefficient is given by its empirical counterpart known as the Pearson correlation coefficient (PCC). We would like to warn against the intentional use of highly heterogeneous data sets in the analysis of regulatory relationships among genes, even if such relationships are perceived as merely statistical associations and not causal effects. The most widely used approach to inferring genetic regulatory structures is to collect microarray data from different sources of tissues (sometimes even from different species) and identify coexpressed genes from this mixed set of data treating it as a sample in the statistical sense, i.e., as a collection of i.i.d. random vectors. For example, one specific data set of this type includes 101 samples from 43 different human tissues and three cell lines.

Some of such observations may be replicated, i.e., represented by arrays obtained from different subjects (usually in small numbers), and some may be represented by only a single array. The well-known Novartis Gene Atlas represents one of the most extreme examples with only one array per each tissue type. The strength of association of gene expression levels is frequently measured by simply calculating the PCC from the pooled data set as recommended by the originators of the relevance network concept. Even if every group (tissue type) includes many arrays, such heterogeneous data sets are not amenable to correlation analysis. This follows from the fact that the compounded correlation coefficient, i.e., the population characteristic ρ_c to which the PCC converges in large samples, is a function of many parameters such as the within-group first and second moments of the marginal distributions, covariances, and expected proportions of each group in the mixture. These parameters are not uniquely determined by ρ_c and the PCC can no longer be thought of as an estimator for a clearly defined measure of dependence. In particular, the hypothesis $H_0 : \rho_c = 0$ becomes meaningless from the statistical standpoint.

2. An Alternative Representation of ρ(*X*, *Y*) and Its Implications

Recalling the model given by (1), we derive a formula that allows us to better understand the principal difficulty brought about by the random nature of the parameter ν. Let us find the covariance between the unobservable r.v.s $\frac{1}{\nu}U$ and $\frac{1}{\nu}V$. We have

$$\begin{aligned}\mathrm{Cov}\left(\frac{1}{\nu}U, \frac{1}{\nu}V\right) &= \mathrm{IE}\left(\frac{1}{\nu}\sum_{s=1}^{\nu} X_s \frac{1}{\nu}\sum_{k=1}^{\nu} Y_k\right) - \mathrm{IE}\frac{1}{\nu}\sum_{s=1}^{\nu} X_s \mathrm{IE}\frac{1}{\nu}\sum_{k=1}^{\nu} Y_k \\ &= \sum_{n=1}^{\infty}\frac{1}{n^2}\sum_{s=1}^{n}\sum_{k=1}^{n}\mathrm{IE}(X_s Y_k)\mathrm{IP}\{\nu = n\} \\ &\quad - \sum_{n=1}^{\infty}\frac{1}{n}\sum_{s=1}^{n}\mathrm{IE}X_s\mathrm{IP}\{\nu = n\}\sum_{n=1}^{\infty}\frac{1}{n}\sum_{k=1}^{n}\mathrm{IE}Y_k\mathrm{IP}\{\nu = n\}\end{aligned} \tag{5}$$

where IP and IE are the symbols of probability and expectation, respectively.

Consider the first term

$$\begin{aligned}\sum_{n=1}^{\infty}\frac{1}{n^2}\sum_{s=1}^{n}\sum_{k=1}^{n}\mathrm{IE}(X_s Y_k)\mathrm{IP}\{\nu = n\} &= \sum_{n=1}^{\infty}\frac{1}{n^2}\sum_{s=1}^{n}\sum_{k=1}^{n}\mathrm{IE}X_s\mathrm{IE}Y_k\mathrm{IP}\{\nu = n\} \\ &\quad + \sum_{n=1}^{\infty}\frac{1}{n^2}\sum_{s=1}^{n}\mathrm{Cov}(X_s, Y_s)\mathrm{IP}\{\nu = n\} \\ &= \mathrm{IE}X\mathrm{IE}Y + \mathrm{Cov}(X, Y)\sum_{n=1}^{\infty}\frac{1}{n}\mathrm{IP}\{\nu = n\}.\end{aligned}$$

For the second term in (5), we have

$$\sum_{n=1}^{\infty}\frac{1}{n}\sum_{s=1}^{n}\mathrm{IE}X_s\mathrm{IP}\{\nu = n\}\sum_{n=1}^{\infty}\frac{1}{n}\sum_{k=1}^{n}\mathrm{IE}Y_k\mathrm{IP}\{\nu = n\} = \mathrm{IE}X\mathrm{IE}Y.$$

It follows from formula (5) that

$$\mathrm{Cov}\left(\frac{1}{\nu}U, \frac{1}{\nu}V\right) = C \cdot \mathrm{Cov}(X, Y)$$

where

$$C = \sum_{n=1}^{\infty}\frac{1}{n}\mathrm{IP}\{\nu = n\} = \mathrm{IE}\frac{1}{\nu}$$

is an unknown constant. Considering the variances of U/ν and V/ν in a similar way, we arrive at the following important assertion:

$$\rho\left(\frac{1}{\nu}\sum_{i=1}^{\nu} X_i, \frac{1}{\nu}\sum_{i=1}^{\nu} \Upsilon_i\right) = \rho(X, \Upsilon) \quad (6)$$

This formula implies that estimating the correlation between the unobservable variables X and Υ in each gene pair amounts to estimating the correlation between their averages over a random number of cells, thereby showing the earlier-mentioned nonidentifiability aspect of the problem in terms of the basic random variables. Note that the model given by (1) can be represented as

$$U = \nu\left(\frac{1}{\nu}\sum_{i=1}^{\nu} X_i\right) = \nu\bar{X}, \quad V = \nu\left(\frac{1}{\nu}\sum_{i=1}^{\nu} \Upsilon_i\right) = \nu\bar{\Upsilon},$$

where the correlation between $\bar{X}$ and $\bar{\Upsilon}$ is the same as that between X and Υ, albeit the distributions of the corresponding vectors can be arbitrarily dissimilar. The above representation shows that the r.v. ν can be interpreted as a multiplicative random noise as long as the main focus is on pairwise correlations.

The noise ν and the signals $\bar{X}$ and $\bar{\Upsilon}$ are inherently dependent under this model. Therefore, the popular model of independent random effect is unlikely to serve a good approximation to the aggregated signals. In Subheading 6, we will invoke formula (6) in our discussion of the utility of the Law of Large Numbers within the framework of model (1).

Formula (6) also illustrates one restrictive assumption behind the model that may have gone unnoticed in its construction. Specifically, the assumption that (X_i, Υ_i) are i.i.d. random vectors implies exchangeability of these vectors across cells and subjects so that the joint distribution of (X, Υ) exhaustively describes both types of variability in formula (1).

3. Assessing the Effect of Signal Aggregation

While our discussion at the end of the previous section suggests that model (1) is quite simplistic, we presently have no better vehicle to assess the potential deviation of the correlation between X and Υ from that between U and V. To gain an idea of how strong the effect of the parameter ν variability can be, let us first compute the coefficient $R = \rho(U, V)$ for some parameter values, assuming that gene expressions within single cells are stochastically independent ($\rho(X, \Upsilon) = 0$). By way of example, suppose $\sigma_\nu / \mu_\nu = 0.23$ and $\mu_\nu = 2 \times 10^5$ cells. From formula (3), we obtain $R = 0.999942$ for

$a=1$, $b=2$ and $R=0.999952$ for $a=1$, $b=5$. When setting $\rho=0.5$ or $\rho=0.9$, the values of R change only in the fifth digit. The same magnitude of R still stands even when $\rho=-0.9$. Notwithstanding arbitrariness of the chosen parameters, this indicates an extremely serious problem arising in studies of dependence structures in general and regulatory networks in particular.

Do our calculations imply that the true correlations between gene expressions are absent or weak? The answer is definitely "No" for the following three reasons. First, the assumption of gene independence is biologically implausible and in conflict with a large body of independent experimental evidence, including the known effects of noncoding RNAs and involvement of genes in biochemical pathways. Second, the situation observed in real data is not as severe as in our sample computations: positive correlations tend to be lower and even a small proportion of negative correlations has been documented. It would appear reasonable that many strong negative correlations are hidden in the prevailing positive correlation structure of microarray data. Third, the unobservable parameters chosen in our computations may be very far from reality. Therefore, we have to base our assessment on real gene expression data rather than imaginary parameters of the model.

4. Remark

It should be noted that negative correlations are typically much more prevalent in normalized versus not normalized data. This does not mean, however, that the commonly used normalization procedures can restore the true correlations. A profound effect of such procedures on the correlation structure of microarray data is well documented. This effect is hardly beneficial as normalization procedures distort the aggregated signals in an unpredictable way and interfere in the true correlation structure. There are also other theoretical reasons for the fact that data normalization does not provide a satisfactory solution to the problem.

From formula (2) it follows that

$$\rho(X,\Upsilon)=\frac{\rho(U,V)\sigma_u\sigma_v - z_v^2\mu_u\mu_v}{\sqrt{(\sigma_u^2 - z_v^2\mu_u^2)(\sigma_v^2 - z_v^2\mu_v^2)}}, \tag{7}$$

where $z_v = \sigma_v / \mu_v$. As a function of z, the coefficient $\rho(X,\Upsilon)$ either decreases monotonically or attains a maximum at the point

$$z^* = \frac{\sqrt{2ab - a^2R - b^2R}}{\sqrt{a^3b + ab^3 - 2a^2b^2R}}, \tag{8}$$

where

$$R = \rho(U,V),\ a = \frac{\mu_u}{\sigma_u},\ b = \frac{\mu_v}{\sigma_v}.$$

Therefore, the effect of signal aggregation is not unidirectional—the correlation coefficient $\rho(X,Y)$ may be smaller or higher than the observed coefficient $\rho(U,V)$. Formula (7) can be represented in a more concise form

$$\rho(X,Y) = \frac{\rho(U,V)\xi_u\xi_v - z_v^2}{\sqrt{(\xi_u^2 - z_v^2)(\xi_v^2 - z_v^2)}}, \tag{9}$$

where $\xi_u = \sigma_u / \mu_u$, $\xi_v = \sigma_v / \mu_v$ are the corresponding variation coefficients.

All the parameters entering formulas (7) or (9) can be estimated from microarray data except for z_v, which is unobservable. However, there are natural mathematical constraints that must be imposed on z_v. First of all, we have to require that $z_v < \xi_u$ for any gene, i.e.,

$$z_v < \min_{1 \le j \le m} \xi_{\mu_j}, \tag{10}$$

where ξ_{u_j}, $j = 1,\ldots,m$, is the variation coefficient for the *j*th gene and *m* is the total number of genes. However, condition (10) does not ensure that $|\rho(X,Y)| \le 1$. To meet the second condition, we derive from (7) the following requirement:

$$z_v^2 \le \frac{\sigma_u^2\sigma_v^2[1 - \rho^2(U,V)]}{\mathrm{Var}(\mu_u V - \mu_v U)}, \tag{11}$$

for all pairs of genes.

The above conditions allow us to deduce a realistic range of possible values of the unobservable variation coefficient *z* from a specific set of microarray data.

If $\rho(X,Y)$ appears to be a monotonically decreasing function of z_v, which property can be verified with real data, then we can use formula (7) to estimate its maximal deviation from $\rho(U,V)$ by evaluating $\rho(X,Y)$ at the right extreme of z_v yielded by conditions (10) and (11). In this case, we obtain a reasonably realistic upper estimate of the actual effect of signal aggregation in accordance with model (1). If $\rho(X,Y)$ passes through a maximum as a function of z_v, this estimate will become conservative to shifts towards lower values of the true correlation coefficients. Such estimates need to be produced for all gene pairs, of course. More accurate estimates of the effect in both directions (up and down) can be obtained by evaluating the behavior of $\rho(X,Y)$ over the whole range of admissible values of z_v in each gene pair, but this approach is computationally extremely expensive and requires parallel computations.

Table 1
Variation coefficients of gene expression levels estimated from different data sets. The average and minimal (across genes) values are presented

Dataset	Average ξ_u	$\min \xi_u$
TELL	0.233	0.188
HYPERDIP	0.267	0.211
BCC	0.299	0.173
PCNORM	0.299	0.213
PCTUM	0.251	0.152

The mean and minimal (across genes) variation coefficients of gene expression were estimated from the following five sets of microarray data:

BCC: Breast cancer cells cultured in vitro (represented solely by "vehicle" control samples that were treated with the medium used to solubilize the inhibitor) with HG_U133A Affymetrix Chip used to produce microarray measurements;

TELL and HYPERDIP: two types of childhood leukemia, U95A Affymetrix Chip;

PCTUM: prostate cancer, U95Av2 Affymetrix Chip;

PCNORM: normal prostate tissue obtained from prostate cancer patients, U95Av2 Affymetrix Chip.

The results are shown in the following Table 1.

These estimates are consistent with the earlier reported observation that the variation coefficients of gene expression are virtually constant across genes. Using the above-described approach, we analyzed all gene pairs in the HYPERDYP data set reporting expression levels of $m = 7084$ genes for $n = 88$ patients with a specific type of childhood leukemia. In this case, Table 1 offered $\min \xi_u^2 = 0.044$ as an upper bound for z_v^2.

A more accurate estimate of 0.041 was given by inequality (11). Therefore, we used the latter value as the conservative estimate of z_v^2 when computing the correlation coefficient $\rho(X,\Upsilon)$ by formula (7). Testing for monotonicity was performed by partitioning the admissible range of z_v^2 (given by condition (11)) into four intervals and using formula (7) to compute the corresponding increments of $\rho(X,\Upsilon)$ for each interval. If at least one increment happened to be positive in a given pair, this event was recorded as a "monotonicity violation". There were less than 0.2 % of all gene pairs that could be suspected for such violations in the HYPERDYP data. While this frequency of monotonicity violation may be considered as quite small, it should be kept in mind that possible shifts in $\rho(X,\Upsilon)$ towards values higher than the observed $\rho(U,V)$ were entirely ignored in this analysis.

5. Signal Aggregation and Technical Noise

Our estimates in previous Table and those resulted from condition (11) give only a rough idea of the magnitude of σ_v / μ and making them more accurate is highly desirable. We discuss one possibility to attain these ends in the present section. Consider an experimental design that supposedly eliminates the biological variation, thereby yielding the information on measurement errors only. Suppose that a sample of n arrays is available that consists solely of technical replicates representing gene expression measurements taken from one and the same subject. Proceeding from the traditional multiplicative noise model,

$$\tilde{X}_j = \varepsilon_j X_j, \quad j = 1,\ldots,m \tag{12}$$

where m is the total number of genes (probe sets), $\tilde{X}_j$ is the observed random signal, and ε_j is an independent random technical (both gene- and array-specific) noise, one would model this situation as

$$\tilde{X}_j = \varepsilon_j C_j, \quad j = 1,\ldots,m \tag{13}$$

where C_j are nonrandom constants.

If the expression levels are log-transformed, we have

$$\log \tilde{X}_j = \log \varepsilon_j + \log C_j.$$

Therefore,

$$\mathrm{Var}(\log \tilde{X}_j) = \mathrm{Var}(\log \varepsilon_j),$$

so that, relying on model (13), one can measure the variance, $\mathrm{Var}(\log \varepsilon_j)$, of the log-transformed technical noise directly from technical replicates. In particular, one can estimate the variance of the mean noise across all probe-sets, i.e.,

$$\sigma_{\bar{\varepsilon}}^2 = \mathrm{Var}\left\{\frac{1}{m}\sum_{j=1}^{m} \log \varepsilon_j\right\}.$$

We resorted to the above line of reasoning in (2) when reanalyzing the Microarray Quality Control Study (MAQC).

For this data set, the estimated $\sigma_{\bar{\varepsilon}}$ is equal to 0.09, which is slightly smaller than the mean (across probe-sets) of estimated standard deviations reported in Klebanov and Yakovlev (2). Since the overwhelming majority of genes have typically much larger (>0.3) standard deviations of their log-expression signals in biological replicates (different subjects), this level of technical noise can be deemed negligibly small. This estimate also leads us to conclude that the true correlation between the unobservable signals $\log X_j$ is really strong. Indeed, the contribution of $\mathrm{Var}\left\{\frac{1}{m}\sum_{j=1}^{m}\log X_j\right\}$ to the variance of log-expressions observed in biological data is much larger than the contribution of $\mathrm{Var}\left\{\frac{1}{m}\sum_{j=1}^{m}\log \varepsilon_j\right\}$ estimated independently from the MAQC data, while a strong correlation between true biological signals (i.e., their values in the absence of measurement errors) is the only explanation for such a discrepancy when the number m of genes is very large. This also explains why the Law of Large Numbers (LLN) is not met in microarray data when applied to log-expression levels across genes.

The situation is no longer the same when we proceed from model (1) in an effort to measure the technical noise stemming from the random nature of the parameter ν. For any gene j, formula (1) gives

$$U_j = \sum_{i=1}^{\nu} X_{ij}, \quad j = 1,\dots,m \tag{14}$$

and it is the parameter ν that plays the role of the technical noise here. It is clear from (14) that the biological variability cannot be entirely eliminated from gene expression signals even when they are produced by purely technical replicates. Designed to assess the technical variability, the experiment described above may only reduce the variance of the r.v.s X_{ij} by eliminating the inter-subject variability, but there will always be some residual biological variability associated with different cells.

Under such experimental conditions, we have

$$\tilde{U}_j = \sum_{i=1}^{\nu} \tilde{X}_{ij}, \quad j = 1,\dots,m \tag{15}$$

where $\tilde{X}_{ij}$ are i.i.d. r.v.s representing the expression levels of the jth gene in different cells obtained from the same subject and their common (conditional) variance is expected to be smaller than that of X_{ij} in (14). Formula (15) also suggests that the MAQC data are far from ideal for the purposes of noise assessment because the technical replicates in this study were produced from a mix of many dissimilar tissue sources; this heterogeneity of samples may inflate the variance of $\tilde{X}_{ij}$ while it should be kept as low as possible.

To remove the scaling factor C_j from the model (13), when deriving the variance of its noise component, we log-transformed the observed expression signals $\tilde{X}_j$. This trick does not work for model (15) and this significantly complicates the noise assessment.

6. The Law of Large Numbers and Random Summation

The following claims seem to be natural in the context of the model given by formulas (1):

1. The observed expression signal U is a result of summation of the inter-cellular signals Xi over a random number of cells ν, thereby defining the basic model structure represented by formulas (1). The random summands Xi are i.i.d. positive r.v.s. independent of ν.
2. While the r.v. ν is nondegenerate, it tends to take on large values with high probability because the number of cells is expected to be large.

In what follows, we examine some indirect corroborative evidence for the above claims.

Suppose for a moment that the number of summands $\nu = k$ is nonrandom. Then the distribution of the corresponding sum in (1) is M-divisible, i.e., it can be represented as the convolution of M distribution functions. In this particular case, the fouth central moment $\mu_4(U)$satisfies the inequality:

$$\mu_4(U) \geq \left(3 - \frac{2}{M}\right)\sigma_u^4. \tag{16}$$

For infinitely divisible distributions, the condition (16) assumes the form

$$\mu_4(U) \geq 3\sigma_u^4. \tag{17}$$

Under mild conditions, these inequalities hold in the case of random ν as well.

If the inequality (16) is met in real biological data, this fact will lend additional support to the presence of signal summation in microarray technology. When testing the corresponding inequalities for empirical counterparts of the moments $\mu_4(U)$ and σ_u in (16) and (17), we observed the event of their violation to be of relatively rare occurrence. For example, the inequality (17) was violated for 18.6 % of the 7,084 genes in the HYPERDIP data. As expected, this proportion was lower for any finite M in (16). Although there is no objective criterion for declaring this frequency consistent with the property of infinite divisibility, we deem it quite low in view of the fact that $\mu_4(U)$ and σ_u in (17) were replaced with their sample counterparts. To corroborate our perception, we generated 7,000 independent samples of size $n = 88$ from a log-normal distribution with parameters $1\mathrm{E}(\log U) = 0.7$ and $\mathrm{Var}(\log U) = 0.09$. The experiment was repeated 1,000 times. The mean proportion of "inconsistent" cases was equal to 23.3 %, suggesting that the random chance of the event under observation may be high even when the underlying distribution is known to be infinitely divisible.

Since the number of cells ν is expected to be large, it is tempting to apply the Law of Large Numbers (LLN) to the normalized random sum

$$Z_k = \frac{1}{\nu_k}\sum_{i=1}^{\nu_k} X_i, \tag{18}$$

where $k \in N$, and make some predictions based on its behavior as $\nu_k \to \infty$ $(k \to \infty)$ in probability. As before, we will assume that the sequence of nonnegative integer-valued r.v.s ν_k is independent of X_i, $i \geq 1$ and $\nu_k \to \infty$ (in probability) as $k \to \infty$. The continuous r.v.s X_i are i.i.d. and positive. If μ_x is finite, it is known that $Z_k \to \mu_x$ as $\nu_k \to \infty$, with both limit relations holding in probability as $k \to \infty$. This is the LLN for random sums.

There is no way of ascertaining whether the LLN is met in real microarray data because the r.v. ν is unobservable. However, we intend to use this powerful tool to predict certain properties of expression signals and then verify them with real data. In doing so, we rely on the following simple result.

7. Assertion

Under the above conditions, the random vector $\mathbf{Z}_k = Z_{1k}, \ldots, Z_{mk}$, with its components defined by

$$Z_{jk} = \frac{1}{\nu_k}\sum_{i=1}^{\nu_k} X_i^{(j)} = \frac{U_k^{(j)}}{\nu_k}, \quad j = 1, \ldots, m, \tag{19}$$

converges in distribution $(\xrightarrow{d})$ to a degenerate random vector as $k \to \infty$.

Now we are in a position to make and test the following two predictions:

Prediction 1

The ratios of the observed expression levels U_j and U_r, $j \neq r$, where $j, r = 1, \ldots, m$ and m is the total number of genes, tend to have small variances. The covariance between different ratios U_j / U_r is expected to be small as well.

Indeed, proceeding from the LLN, we expect the asymptotic relation

$$\frac{U_j}{U_r} = \frac{\frac{1}{\nu}\sum_{i=1}^{\nu} X_{ij}}{\frac{1}{\nu}\sum_{i=1}^{\nu} X_{ir}} \sim \frac{\mathrm{IEX}_{1j}}{\mathrm{IEX}_{1r}}, \quad j \neq r, \tag{20}$$

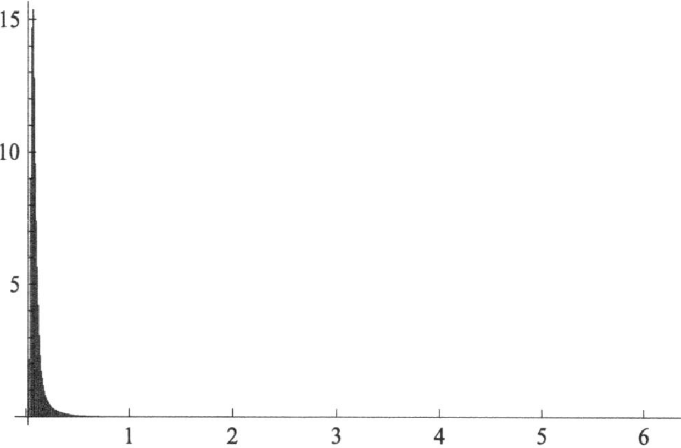

Fig. 1. Histogram of the standard deviations of the the ratios U_j / U_r $(j \neq r)$ in the HYPERDIP data.

to hold true as $\nu \to \infty$ in probability. This suggests that every ratio U_j / U_r is virtually constant (across arrays). The above-proven assertion also suggests that every two ratios of the form: U_j / U_r and U_l / U_q (with different indices) have small covariances.

To verify Prediction 1, we formed all pairs from 1,000 randomly selected genes. The mean (over the gene pairs) standard deviation of the ratios U_j / U_r $(j \neq r)$ in the HYPERDIP data was equal to 0.102, which value is very small compared to the corresponding mean of the estimated expectations $\mathbb{E} U_j / U_r$, the latter value being equal to 1.044.

The histogram of the standard deviations in Fig. 1 illustrates this point further. Shown in Fig. 2 below is the histogram of the estimated covariances between U_j / U_r and U_l / U_q in all quadruples formed from 100 randomly selected genes in the HYPERDIP data set. It is clear that they tend to be small as well. This observation explains the most salient properties of the so-called δ-sequence, as well as a remarkable success of significance testing for differential expression of genes when the relevant methods are applied to the elements of this sequence rather than to the original expression levels.

Prediction 2

The average (expectation) of the ratio U_j / U_r is approximately equal to the ratio of the averages of U_j and U_r, $j \neq r$.

Invoking the LLN, we can assert that

$$\mathbb{E} U_j = \mathbb{E}\nu \mathbb{E} X_{1j}, \quad \mathbb{E} U_r = \mathbb{E}\nu \mathbb{E} X_{1r}$$

Proceeding from the representation

$$\frac{U_j}{U_r} = \frac{\frac{1}{\nu}\sum_{i=1}^{\nu} X_{ij}}{\frac{1}{\nu}\sum_{i=1}^{\nu} X_{ir}},$$

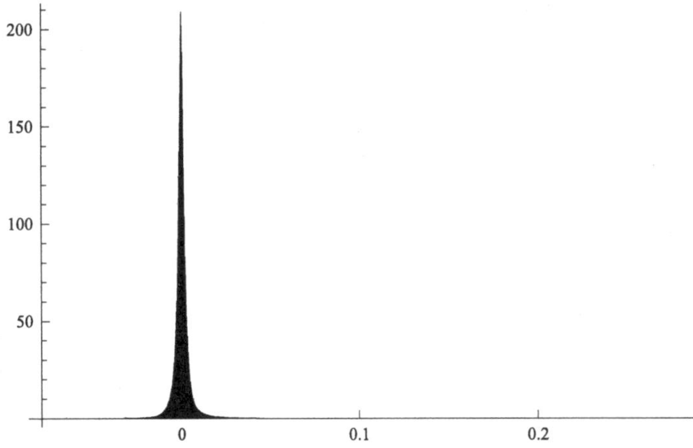

Fig. 2. Histogram of covariances between the ratios of gene expressions in all quadruples from a subset of 100 genes. The HYPERDIP data set.

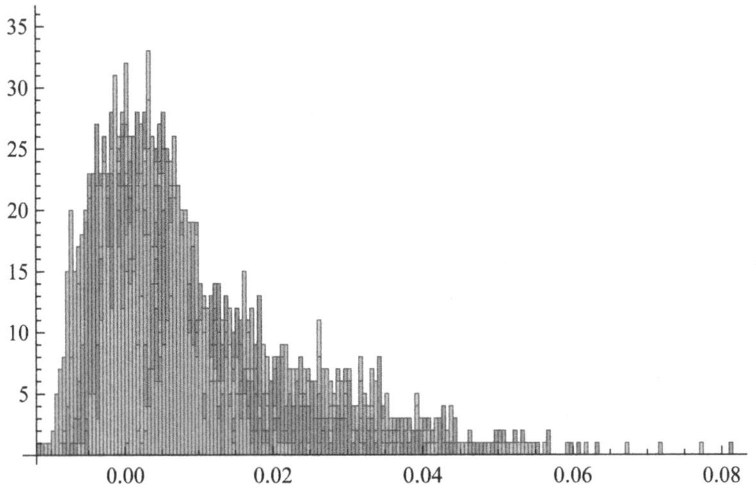

Fig. 3. Histogram of the differences between $E(U_j/U_r)$ and $E(U_j)/E(U_r)$ estimated by replacing the expected values with the corresponding sample means. All pairs from 250 genes. The HYPERDIP data set.

we can claim that

$$\mathrm{E}\left(\frac{U_j}{U_r}\right) \approx \frac{\mathrm{E}U_j}{\mathrm{E}U_r} \tag{21}$$

whenever ν is large with high probability.

Replacing the expected values with their sample counterparts, we computed the absolute difference between the left and right hand sides of the equality (21) for all gene pairs formed from 1,000 randomly selected genes in the HYPERDIP data set. The resultant histogram (Fig. 3 below) clearly indicates that such differences are very small with the mean (across the gene pairs) being equal to 0.006.

Hence, both predictions appear to be consistent with real data. Similar analyses of the other data sets have confirmed this conjecture.

8. Concluding Remarks

A genetic pathway is the set of interactions occurring between a group of genes that depend on each other's individual functions in order to make the aggregate function of the network available to the cell. It is interesting to study such groups of genes as a multivariate vector, and compare such vectors for different states.

We would like to note, that an interesting study in this direction was recently conducted by Galina V. Glazko and Frank Emmert-Streib in "Unite and conquer: univariate and multivariate approaches for finding differentially expressed gene sets," Bioinformatics, vol. 25, 18, 2009, 2348–2354. The authors provide simulations for comparison of four multivariate tests, including one based on *N*-distance. This last appears to be the best almost in all situations. The authors gave also applications to real data.

Acknowledgement

The study was supported by Grant MSM 0021620839 of the Ministry of Education, Czech Republic.

References

1. Chu T, Glymour C, Scheines R, Spirtes P (2003) A statistical problem for inference to regulatory structure from associations of gene expression measurements with microarrays. Bioinformatics 19:1147–1152
2. Klebanov L, Yakovlev A (2007) How high is the level of technical noise in microarray data? Biol Direct 2:9

Chapter 12

Test for Normality of the Gene Expression Data

Bobosharif Shokirov

Abstract

One of the main issues in statistical analysis of gene expression data is testing levels of differentially expressed genes. There are different approaches to address this issue. If in one case, given two probe sets of gene expression levels, we test whether they are differentially expressed or not, in other cases we just check whether the given sample is drawn from the prescribed distribution or not. Assuming that the prescribed distribution is normal we would like to test normality of gene expression data.

Key words: Gene Expression Data, Normality tests

1. Introduction

Recent years have seen a growing interest in the statistical analysis of gene expression data. If in one case we test for differentially expressed levels in two probe sets of genes, in other cases we would like to know whether the given sample is drawn from a prescribed distribution or not. In this work without specifying the parameters, we assume that the prescribed distribution is normal and would like to test whether the sample of the gene expression data are drawn from normal distribution or not. Before we start our discussion, we have to mention at least two main difficulties, which may arise in the statistical analysis of the gene expression data in general, and in testing for normality in particular. These two main problems are:

1. It is known that gene expression data are highly correlated.
2. The number data of gene expression levels is much higher than the number of available observations (thousands against tens).

Andrei Y. Yakovlev et al. (eds.), *Statistical Methods for Microarray Data Analysis: Methods and Protocols*, Methods in Molecular Biology, vol. 972, DOI 10.1007/978-1-60327-337-4_12, © Springer Science+Business Media New York 2013

There are several approaches to overcome the first difficulty (for example, by using δ sequences) but in fact there is no way to improve the second difficulty, to increase the number of observations. Taking into account these two problems, we proceed normality test of the gene expression data in such way that having not enough observations, meaning that the number of observations is not comparable to number of gene expressions, would not matter that much.

2. Some Characterization Theorems of the Normal Distribution

We will need some results of characterization of the normal density function.

1. *Zinger A.* (see ref. 1). Let $x_1, x_2, \ldots, x_n$ be n independent identically distributed random variables. Consider statistics

$$z_1 = \frac{x_1 - \overline{x}}{S}, z_2 = \frac{x_2 - \overline{x}}{S}, \ldots, z_n = \frac{x_n - \overline{x}}{S}, \quad (1)$$

where

$$n\overline{x} = x_1 + x_2 \ldots + x_n, \quad s^2 = \sum_{i=1}^{n} (x_i - \overline{x})^2.$$

Then the random vector $Z = (z_1, z_2, \ldots, z_n)$ is distributed on $n-2$-dimensional sphere

$$S^{n-2} = \begin{Bmatrix} z_1 + z_2 + \ldots + z_n = 0 \\ z_1^2 + z_2^2 + \ldots + z_n^2 = 1 \end{Bmatrix}.$$

It is clear that the distribution of random variables $x_1, x_2, \ldots, x_n$ defines the distribution of the random vector **Z**. Particularly, if random variables $x_1, x_2, \ldots, x_n$ have normal distribution then the random vector **Z** has uniform distribution on the sphere. The inverse problem was solved by Zinger A (1) for the case when $n \geq 6$. Namely, the following theorem was proved by him.

Theorem 1:

If random vector Z has uniformly distribution on the sphere S^{n-2} and $n \geq 6$, then random variables $x_1, x_2, \ldots, x_n$ have normal distribution.

Theorem 1 allows to replace normality test by testing uniformity of the given sample.

2. *Sakata T.* (see ref. 2). Let $(X_{i1}, X_{i2}, \ldots, X_{i2k}), i = 1, \ldots, n$ be a sequence of samples with equal size $2k$, being independently drawn from a population Π_i with a continuous density function

$$\frac{1}{\sigma_i} p\left(\frac{x - \mu_i}{\sigma_i}\right),$$

where $p(.)$ is a symmetric density and $-\infty < \mu_i < +\infty$, $\sigma_i > 0$, $(i = 1,\ldots,n)$. Sakata studied the problem of testing hypothesis about the common density $p(\cdot)$ where the unknown parameters (μ_i, σ_i) may change from one population to another, and the sample size $2k$, is so small that the estimate of the unknown parameters with enough accuracy are not available. Therefore he proposes a series of transformation to eliminate the unknown parameters (μ_i, σ_i), which could be expressed as following

$$Z_i = \frac{|Y_i|}{\sqrt{\sum_{i=1}^{k} Y_i^2}} \text{with } Y_i = X_{2i-1} - X_{2i}, \quad i = 1,\ldots,k. \tag{2}$$

The following statements were proved by Sakata (2, 3).

Theorem 2:

1. *Let $h(Z)$ be the density function of the statistic $Z = (Z_1, Z_2, \ldots, Z_k)$, which takes value on $S_+^{k-1} = \left\{\sum_{i=1}^{k} Z_i^2 = 1, Z_i \geq 0, i = 1,\ldots,k\right\}$*

Then

$$h(Z) = c\int_0^{\infty} s^{k/2-1} f(\sqrt{sZ_1})\ldots f(\sqrt{sZ_k})ds, \tag{3}$$

where $f(\cdot)$ is convolution of $p(\cdot)$ and c is a constant.

2. *If $p(.)$ is the standard normal density, then the random variable $Z = (Z_1, Z_2, \ldots, Z_k)$, is uniformly distributed on S_+^{k-1}.*

3. *Let $p(.)$ have a differentiable bounded convolution. If for $k \geq 3$ the following statistic $Z = (Z_1, Z_2, \ldots, Z_k)$, is uniformly distributed on S_+^{k-1} then $p(.)$ is the density of the standard normal distribution.*

3. Test of Uniformity on the Sphere by Using N-Distances

Theorems of Zinger and Sakata show that if the transformed sample has uniform distribution of the sphere, then the original sample has normal distribution (without specifying the parameters of the normal distribution). Therefore, it is enough to test uniformity of the resulted sample. For this goal we use a test statistics, generated by N-distances (4, 5).

Let (X, M) be a measurable space and B the set of all probability measures on it. Suppose that $\mathcal{L}$ is a strongly negative definite kernel on χ^2, and denote by $B_{\mathcal{L}}$ the set of all probability measures μ under condition

$$\int_{\chi}\int_{\chi} \mathcal{L}(x, y)d\mu(x)d\mu(y) < \infty.$$

Introduce the following distance (see ref. 4, 5):

$$N(\mu,\nu)=\left(2\int_{\chi}\int_{\chi}L(x,y)d\mu(x)d\nu(y)-\int_{\chi}\int_{\chi}L(x,y)d\mu(x)d\mu(y)\right.$$
$$\left.-\int_{\chi}\int_{\chi}L(x,y)d\nu(x)d\nu(y)\right)^{1/2}.$$

As examples of strongly negative definite kernels one can mention

$$\mathcal{L}(x,y)=\|x-y\|,$$
$$\mathcal{L}(x,y)=\frac{\|x-y\|}{1+\|x-y\|},$$
$$\mathcal{L}(x,y)=\log(1+\|x-y\|^2)$$

for the case of $\chi=R^d$. The same kernels can be used for the case of the $d-1$—dimensional sphere.

4. Test Procedure

Now we proceed the test of normality for gene expression data. Let m be number of genes and n sample size of the gene expression data. Then the matrix $X=(X_{ij})$, $i=1,\ldots,m$ $j=1,\ldots,n$ denotes n observations of the gene expression levels, where element X_{ij} is the j-th observed level of the i-th gene. We proceed testing normality for each gene $i=1,\ldots,m$. Following Sakata, from the i-th raw of the matrix $\mathbf{X}$ we obtain the sample $\Upsilon_{i1},\Upsilon_{i2},\ldots,\Upsilon_{ik}$, where $\Upsilon_{ij}=X_{i2j-1}-X_{i2j}$, $j=1,\ldots,k$ with $k=n/2$. For the odd number n we skip the last observation.
Using transformation Eq. 2 one can obtain

$$Z_{ij}=\frac{|\Upsilon_{ij}|}{\sqrt{\sum_{j=1}^{k}\Upsilon_{ij}^2}},\quad i=1,\ldots.m,\quad j=1,\ldots,k. \tag{4}$$

Applying spherical coordinates

$$\Upsilon_{i1}=\rho\cos\varphi_1$$
$$\Upsilon_{i2}=\rho\sin\varphi_1\cos\varphi_2$$
$$\Upsilon_{i3}=\rho\sin\varphi_1\sin\varphi_2\cos\varphi_3$$
$$\ldots\ldots\ldots\ldots$$
$$\Upsilon_{ik-1}=\rho\sin\varphi_1\ldots\sin\varphi_{k-2}\cos\varphi_{k-1}$$
$$\Upsilon_{ik}=\rho\sin\varphi_1\ldots\sin\varphi_{k-2}\sin\varphi_{k-1}$$

we have

$$\rho Z_{ij} = \Upsilon_{ij}, \quad j = 1,\ldots,k. \tag{5}$$

4.1. Case k = 2

1. Let $k = 2$ and for every $i = 1,\ldots,m$ let $f(y_1)f(y_2)$ be joint density of the variates $(\Upsilon_{i1}, \Upsilon_{i2})$. The change to spherical coordinates gives

$$f(y_1)f(y_2)dy_1dy_2 = f(\rho\cos\phi_1)f(\rho\sin\phi_1)\rho d\rho d\phi_1$$

and the joint density of random variables Z_{i1}, Z_{i2} would be

$$\int_0^\infty f(\rho\cos\varphi_1)f(\rho\sin\varphi_1)\rho d\rho.$$

Let

$$\int_0^\infty f(\rho\cos\varphi_1)f(\rho\sin\varphi_1)\rho d\rho = c, \tag{6}$$

where c is some constant. At this point the question we would like to answer is: does the function $f(.)$ represent the density of the normal distribution. Substitution $r = \rho\cos\varphi_1$ $r = \cos\varphi_1$ in Eq. 6 gives

$$\int_0^\infty f(r)f(r\tan\varphi_1)rdr = c\cos^2\varphi_1, \tag{7}$$

or

$$\int_0^\infty f(r)f(tr)rdr = \frac{c}{1+t^2}, \tag{8}$$

where $t = \tan\varphi_1$. Multiplying both sides of Eq. 8 by t^s and integrating with respect to t over the interval $[0, \backslash\infty)$ we obtain

$$\int_0^\infty\int_0^\infty f(r)f(tr)rdrt^s dt = \int_0^\infty \frac{c}{1+t^2}t^s dt. \tag{9}$$

The inner integral can be transformed in the following way:

$$\int_0^\infty f(tr)t^s dt = \frac{1}{r^{s+1}}\int_0^\infty f(y)y^s dy.$$

Then

$$\int_0^\infty f(r)\frac{1}{r^s}dr\int_0^\infty f(y)y^s dy = \int_0^\infty \frac{c}{1+t^2}t^s dt. \tag{10}$$

Expression

$$\varphi(s) = \int_0^\infty f(y)y^s dy \tag{11}$$

is the Mellin transformation of function $f(x)$. Then we have the following functional equation:

$$\phi(-s)\phi(s) = h(s), \tag{12}$$

where

$$h(s) = c\int_0^\infty \frac{t^s}{1+t^2}dt.$$

But

$$\int_0^\infty \frac{x^s}{1+x^2}dx = \frac{1}{2}\sec\frac{\pi s}{2}.$$

For the normal density

$$\varphi(s) = c2^{\frac{s-1}{2}}\Gamma\left(\frac{1-s}{2}\right).$$

We have

$$\Gamma\left(\frac{1+s}{2}\right)\Gamma\left(\frac{1-s}{2}\right) = \left(\Gamma\left(\frac{1}{2}\right)\right)^2 \frac{1}{\cos\frac{s\pi}{2}}$$

Thus, for the density $f(x)$, which is an even function we have

$$\varphi(s) = \int_0^\infty f(y)y^s dy$$

and the Eq. 12 can be written in the form

$$\varphi(s)\varphi(-s) = c\frac{1}{\cos\frac{\pi s}{2}}.$$

From the identity $\Gamma(1-z)\Gamma(z) = \frac{\pi}{\sin \pi z}$ by changing $z = (1-s)/2$ we get

$$\Gamma\left(\frac{1+s}{2}\right)\Gamma\left(\frac{1-s}{2}\right) = \frac{\pi}{\sin(\pi(1-s)/2)}.$$

If we denote $\varphi(s) = \psi\left(\frac{1-s}{2}\right)$ and $\varphi(-s) = \psi\left(\frac{1+s}{2}\right)$, we have

$$\psi\left(\frac{1-s}{2}\right)\psi\left(\frac{1+s}{2}\right) = \frac{c}{\cos(\pi s/2)}$$

and, finally,

$$g(1-z)g(z) = \frac{\pi}{\sin \pi z},$$

where $z = (1-s)/2$. But it is not necessary $g(z)$ to be Gamma function. From

$$\cos x = \prod_{n=1}^{\infty}\left(1 - \frac{x^2}{\pi^2(n-1/2)^2}\right),$$

we get

$$\cos\frac{\pi s}{2} = \prod_{n=1}^{\infty}\left(1 - \frac{s^2}{(2n-1)^2}\right)$$

and

$$\frac{1}{\cos(\pi s/2)} = \prod_{n=1}^{\infty}\frac{1}{1-(s^2/(2n-1)^2)}.$$

and by different "regrouping" we get

$$\varphi(s)\varphi(-s) = \prod_{n=1}^{\infty}\frac{1}{1-(s^2/(2n-1)^2)}\varphi^2(0),$$

keeping $\varphi(s)$ as a Mellin transformation.

We can formulate all this as the following result.

Theorem 1:
For the case $k = 2$ the result by Sakata is not true, that is there are nonnormal random variables for which **Z** is uniformly distributed over S_+^1.

4.2. The Case of $k = 3$

In this situation the characterization obtained by Sakata holds, and it is possible to use it to construct the test of the normality of gene expression levels. Of course, in view of this characterization, it is enough to test the uniformity of the distribution of the vector **Z** on S_+^2. To this goal we use the Bakshaev test (see ref. 1) generated by $\mathcal{N}$ -distance with corresponding kernel $\mathcal{L}(x,y) = \|x-y\|$. Here we give the results of the testing procedure.

We are given two sets of data HYPERDIP and TELL. HYPERDIP consists of 88 observations and TELL of 79 observations. For both datasets each observation has dimension 7,084, which corresponds to the number on genes. We would like to test normality of these data.

As it was stated above, to test normality of the given data we use a statistic T to test the uniformity of transformed data on the sphere. If as a result of this test we will have uniformly distributed on the sphere data, then by virtue of theorems of A. Zinger and T. Sakata, where connection between normal and uniform random variables are established, original data have normal distribution.

In the meantime we consider observations for each gene as a vector of dimensions 88 and 79 for HYPERDIP and TELL, correspondingly. Below algorithm for testing normality is given.

Algorithm 1

1. Take k vectors $X_i = \{x_{i1},\ldots,x_{in}\}$, $i = 1,\ldots,k,$, where $n = 79(88)$.
2. Construct k vectors and $Y_i = \{y_{i1},\ldots,y_{im}\}$, where $y_{ij} = x_{i,2j} - x_{i,2j-1}$, $i = 1,\ldots,k$, $j = 1,\ldots,m$, $m = [n/2]$, and $[q]$ is integer part of q.
3. Estimate the covariance matrix

$$\Sigma = \begin{pmatrix} S_{y1y1} & S_{y1y2} & \cdots & S_{y1ym} \\ S_{y2y1} & S_{y2y2} & \cdots & S_{y2ym} \\ \cdots & \cdots & \cdots & \cdots \\ S_{ymy1} & S_{ymy2} & \cdots & S_{ymym} \end{pmatrix},$$

where $S_{y_iy_j} = S_{y_jy_i}$ and

$$S_{y_iy_j} = \frac{1}{m-1}\sum_{i=1}^{m}\left[\left(y_i - \frac{1}{m}\sum_{i=1}^{m} y_i\right)\left(y_j - \frac{1}{m}\sum_{j=1}^{m} y_j\right)\right].$$

4. Take the l-th coordinate of vectors Y_i, $i = 1,\ldots,k$ construct k-dimensional vectors $\{y_{1l},\ldots,y_{kl}\}$ for all $l = 1,\ldots,m$ and transform them into k-dimensional vectors $\{z_{1l},\ldots,z_{kl}\}$, $l = 1,\ldots,m$ by

$$\{z_{1l},\ldots,z_{kl}\} = \Sigma^{-1/2}\{y_{1l},\ldots,y_{kl}\}, \quad l = 1,\ldots,m.$$

5. From the first, second, etc. k-th coordinates of the vectors $\{z_{1l},\ldots,z_{kl}\}$ $l = 1,\ldots,m$ construct m-dimensional vectors

$$Z_i = \{z_{i1},\ldots,z_{im}\}, \quad i = 1,\ldots,k.$$

6. Merging vectors $Z_1,\ldots,Z_k$ obtain vector $Z = \{z_1,\ldots,z_{mk}\}$ of dimension mk, where $z_i = z_{1i}, z_{m+i} = z_{2i}, z_{2m+i} = z_{3i},\ldots,$ $z_{(k-1)m+i} = z_{ki}$ for $i = 1,\ldots,m$.
7. Split vector Z by groups of three elements, divide each group by its norm and obtain the matrix $A = (a_{ij})_{M\times 3}$, where

$$a_{ij} = \frac{z_{3(i-1)+j}}{\sqrt{z_{3i-2}^2 + z_{3i-1}^2 + z_{3i}^2}}, \quad i = 1,\ldots,M = mk/3, \quad j = 1,2,3.$$

8. Calculate T statistic as

$$T = M - \frac{3}{2M}\sum_{i=1}^{M-1}\sum_{j=i+1}^{M} \| A_i - A_j \|,$$

where

$$A_i = \left\{ \frac{z_{3i-2}}{\sqrt{z_{3i-2}^2 + z_{3i-1}^2 + z_{3i}^2}}, \frac{z_{3i-1}}{\sqrt{z_{3i-2}^2 + z_{3i-1}^2 + z_{3i}^2}}, \frac{z_{3i}}{\sqrt{z_{3i-2}^2 + z_{3i-1}^2 + z_{3i}^2}} \right\}$$

denotes the i-th raw of the matrix A.

9. By simulation find percentage points of the statistic T, look for percentage point corresponding to the value of T found in 8 and decide to reject hypothesis on uniformity (normality) of the data or not.

Algorithm 2

1. Take k vectors $X_i = \{x_{i1},\ldots,x_{in}\}$, $i = 1,\ldots,k$, where $n = 79(88)$.
2. Construct k vectors and $\Upsilon_i = \{y_{i1},\ldots,y_{im}\}$, where $y_{ij} = x_{i,2j} - x_{i,2j-1}$ $i = 1,\ldots,k,$ $j = 1,\ldots,m,$ $m = [n/2]$.
3. Generate k random vectors $\varepsilon_i^{\sigma^2} = \left\{\varepsilon_{i1}^{\sigma^2},\ldots,\varepsilon_{im}^{\sigma^2}\right\}$, $i = 1,\ldots,k$ of dimension m from normal distribution with zero mean and dispersion $\sigma^2 : 0 \le \sigma^2 \le 1$. Change σ^2 from 0 to 1 by step h. Let $q = [1/h]$ be the number of steps. Then for every step r we obtain vectors $\varepsilon_i^r = \{\varepsilon_{i1}^r,\ldots,\varepsilon_{im}^r\}$, $i = 1,\ldots,k,$ $r = 1,\ldots,q.$
4. For every $r = 1,\ldots,q$ construct vectors $\Upsilon_i^\varepsilon[r] = \{y_{i1}^\varepsilon[r],\ldots,y_{im}^\varepsilon[r]\}$, where $y_{ij}^\varepsilon[r] = y_{ij} + \varepsilon_{ij}^r$, $i = 1,\ldots,k,$ $j = 1,\ldots,m.$
5. Estimate the covariance matrix

$$\sum[r] = \begin{pmatrix} S_{y_1^\varepsilon[r]y_1^\varepsilon[r]} & S_{y_1^\varepsilon[r]y_2^\varepsilon[r]} & \cdots & S_{y_1^\varepsilon[r]y_m^\varepsilon[r]} \\ S_{y_2^\varepsilon[r]y_1^\varepsilon[r]} & S_{y_2^\varepsilon[r]y_2^\varepsilon[r]} & \cdots & S_{y_2^\varepsilon[r]y_m^\varepsilon[r]} \\ \cdots & \cdots & \cdots & \cdots \\ S_{y_m^\varepsilon[r]y_1^\varepsilon[r]} & S_{y_m^\varepsilon[r]y_2^\varepsilon[r]} & \cdots & S_{y_m^\varepsilon[r]y_m^\varepsilon[r]} \end{pmatrix}, \quad r = 1,\ldots,q,$$

where $S_{y_i^\varepsilon[r]y_j^\varepsilon[r]} = S_{y_j^\varepsilon[r]y_i^\varepsilon[r]}$ and

$$S_{y_i^\varepsilon[r]y_j^\varepsilon[r]} = \frac{1}{m-1}\sum_{i=1}^{m}\left[\left(y_i^\varepsilon[r] - \frac{1}{m}\sum_{i=1}^{m} y_i^\varepsilon[r]\right)\left(y_j^\varepsilon[r] - \frac{1}{m}\sum_{j=1}^{m} y_j^\varepsilon[r]\right)\right].$$

6. For every $r = 1,\ldots,q$ take the l-th coordinate of vectors $\Upsilon_i^\varepsilon[r]$, $i = 1,\ldots,k$, construct k-dimensional vectors

$\{y_{1l}^{\varepsilon}[r],\dots,y_{kl}^{\varepsilon}[r]\}$ for all $l=1,\dots,m$ and transform them into k-dimensional vectors $\{z_{1l}[r],\dots,z_{kl}[r]\}$ by

$$\{z_{1l}[r],\dots,z_{kl}[r]\}=\left(\sum[r]\right)^{-1/2}\{y_{1l}^{\varepsilon}[r],\dots,y_{kl}^{\varepsilon}[r]\},\quad l=1,\dots,m.$$

7. For every $r=1,\dots,q$ from the first, second, etc. k-th coordinates of the vectors $\{z_{1l}[r],\dots,z_{kl}[r]\}$, $l=1,\dots,m$ construct m-dimensional

   vectors $Z_i[r]=\{z_{i1}[r],\dots,z_{im}[r]\}$, $i=1,\dots,k$.

8. Merging vectors $Z_1[r],\dots,Z_k[r]$ obtain vector $Z[r]=\{z_1[r],\dots,z_{mk}[r]\}$ of dimension mk, where

   $$z_j[r]=z_{1j}[r],z_{m+j}[r]=z_{2j}[r],z_{2m+j}[r]=z_{3j}[r],\dots,z_{(k-1)m+j}[r]=z_{kj}[r],$$

   for $j=1,\dots,m$.

9. Split vector $Z[r]$ by groups of three elements, divide each group by its norm and obtain the matrix $A[r]=(a_{ij}[r])_{M\times 3}$, where

$$a_{ij}[r]=\frac{z_{3(i-1)+j}[r]}{\sqrt{(z_{3i-2}[r])^2+(z_{3i-1}[r])^2+(z_{3i}[r])^2}},\quad i=1,\dots,M=mk/3,\quad j=1,2,3.$$

10. For every $r=1,\dots,q$ calculate $T[r]$ statistic as

$$T[r]=M-\frac{3}{2M}\sum_{i=1}^{M-1}\sum_{j=i+1}^{M}\|A_i[r]-A_j[r]\|,$$

where

$$A_i[r]=\{a_{i1}[r],a_{i2}[r]a_{i3}[r]\}.$$

11. By simulation find p-value of the statistic $T[r]$ for $r=1,\dots,q$.

Algorithm 3

Algorithm 3 is almost the same as Algorithm 2. The only differences is that we deal with only two vectors and add the same perturbation

$$\varepsilon_i^{\sigma^2}=\left\{\varepsilon_{i1}^{\sigma^2},\dots,\varepsilon_{im}^{\sigma^2}\right\},\quad i=k,l$$

vectors to all pairs $\{\Upsilon_k,\Upsilon_l\}$.

5. Results

In Algorithm 2 we put $\mathrm{k}=2$ and $q=10$. This means that we deal with two n-dimensional vectors X_1 and X_2 and changed σ^2 from 0.1 to 1 by step $h=0.1$. Results for this case are shown in Table 1.

Table 1
Results for $k=2$

Data	HYPERDIP		TELL	
σ^2	*T*-Test	*p*-Value	*T*-Test	*p*-Value
0.1	0.771159	0.60204	0.820451	0.55281
0.2	0.685192	0.69559	0.91904	0.46094
0.3	1.31716	0.2059	0.681358	0.69997
0.4	0.574453	0.8201	0.572117	0.82255
0.5	1.03035	0.37197	0.507783	0.88824
0.6	1.30829	0.20976	0.520114	0.87593
0.7	0.864125	0.51091	1.15109	0.29276
0.8	1.67603	0.09495	1.61273	0.10864
0.9	0.586329	0.8067	1.19367	0.26803
1.	0.721547	0.65526	1.05492	0.35469

As seen from Table 1, the minimum of *p*-values for HYPERDIP and TELL data, correspondingly, are 0.09495 and 0.10864. Therefore we cannot reject uniformity of transformed vectors $Z[r]$ and correspondingly the normality of the vectors X_1 and X_2.

5.1. Results Derived by Slightly Modified Version of Algorithm 2

If in item 3 of Algorithm 2 instead of generating random vectors $\varepsilon_i^{\sigma^2} = \left\{\varepsilon_{i1}^{\sigma^2},\ldots,\varepsilon_{im}^{\sigma^2}\right\}$, $i = 1,\ldots,k$ we generate *m*-dimensional random vectors from *k*-variate normal distribution with zero mean vector and covariance matrix of the form

$$X_i = \begin{pmatrix} \rho_1 & 0 & \ldots & 0 \\ 0 & \rho_2 & \ldots & 0 \\ \ldots & \ldots & \ldots & \ldots \\ 0 & 0 & \ldots & \rho_k \end{pmatrix}$$

and change every ρ_i from 0 to 1 by step *h*, then we get kq^k *m*-dimensional random vectors $\varepsilon_{ij}^{r_j} = \left\{\varepsilon_{i1}^{r_j},\ldots,\varepsilon_{im}^{r_j}\right\}$ where $r_j = 1,\ldots,q$. For $k = 2$ and $q = 10$ values of the statistic $T[r_j]$ and corresponding their *p*-values could be represented in the form of Tables.

As seen from Tables 2, 3, 4, and 5 forTELL data the minimal *p*-value of the T statistic is 0.02912 (and maximal 0.99728).

Table 2
HYPERDIP: Values of the *T* statistics for $k=2$ and $q=10$

ρ_1 / ρ_2	0.1	0.2	0.3	0.4	0.5	0.6	0.7	0.8	0.9	1.
0.1	0.78730	0.76470	1.29100	1.07494	0.705487	0.51914	0.904572	1.17394	0.551013	0.724801
0.2	1.18870	0.60168	0.47037	2.42486	0.808641	1.02354	0.982085	0.544208	1.09511	0.914443
0.3	0.74730	1.02567	0.67016	0.54548	0.544037	0.477699	1.07409	0.888156	0.565794	0.687165
0.4	0.59451	1.06202	0.60046	0.71704	0.75103	0.487955	0.999959	1.22685	0.75088	0.575707
0.5	0.58435	1.33294	1.22582	0.461474	1.16725	0.548474	0.60196	0.733606	0.608818	1.06447
0.6	1.00061	0.60450	1.20465	0.51216	0.909385	0.441835	0.809695	0.660856	0.733815	1.22845
0.7	1.49516	0.33192	1.46233	0.597134	0.589026	0.908223	0.460334	0.524735	0.544499	0.523826
0.8	2.41706	2.72510	0.73522	0.874156	0.807099	1.1679	2.0509	1.0778	1.08673	2.04638
0.9	1.02390	0.97246	1.34208	0.692538	1.23231	0.880708	0.628466	0.860174	0.549219	1.00033
1.	0.55103	0.67688	0.69134	1.75304	1.92841	0.789767	0.750624	1.20052	1.47876	0.768511

Table 3
HYPERDIP: *p*-values of the *T* statistics for $k=2$ and $q=10$

ρ_1 / ρ_2	0.1	0.2	0.3	0.4	0.5	0.6	0.7	0.8	0.9	1.
0.1	0.58558	0.60902	0.21758	0.34067	0.67311	0.87693	0.47365	0.27869	0.84456	0.65183
0.2	0.27063	0.78958	0.92187	0.01788	0.56446	0.37701	0.40896	0.85168	0.32785	0.46505
0.3	0.62736	0.37568	0.71263	0.85032	0.85181	0.91587	0.34128	0.4883	0.82891	0.69297
0.4	0.79732	0.3496	0.79097	0.66014	0.62329	0.90693	0.39488	0.24985	0.62348	0.81873
0.5	0.80918	0.19924	0.25027	0.92891	0.28292	0.84724	0.7892	0.64219	0.78186	0.34783
0.6	0.3943	0.78661	0.26149	0.88378	0.46934	0.94432	0.56328	0.72284	0.64189	0.24909
0.7	0.14092	0.99354	0.15103	0.79473	0.80335	0.47029	0.92971	0.87143	0.85144	0.87235
0.8	0.01828	0.00896	0.64034	0.50159	0.56612	0.28246	0.04131	0.33882	0.33296	0.04173
0.9	0.37681	0.41639	0.19526	0.68709	0.24702	0.49558	0.75881	0.51439	0.84634	0.39451
1.	0.84453	0.70527	0.6884	0.08067	0.05476	0.58332	0.62374	0.26378	0.14602	0.60471

For HYPERDIP data the minimal *p*-value of the *T* statistic is 0.00896 (and maximal 0.99354). Across ρ_1 the mean values of *p*-values are: 0.486491, 0.523389, 0.482984, 0.574578, 0.513618, 0.669995, 0.532482, 0.522197, 0.622531, 0.51388. We cannot reject normality

Table 4
TELL: Values of the *T* statistics for $k=2$ and $q=10$

ρ_1 / ρ_2	0.1	0.2	0.3	0.4	0.5	0.6	0.7	0.8	0.9	1.
0.1	0.903866	0.82169	1.10944	0.66126	0.823747	0.574618	0.646146	0.852312	0.697882	0.539181
0.2	0.396154	1.93175	0.731969	1.04137	1.82724	0.782166	1.06444	1.07859	1.24953	0.714262
0.3	0.894212	0.79538	1.26044	1.59436	0.945505	0.654897	0.403676	1.86983	1.70828	0.98396
0.4	0.592341	0.900685	0.505571	1.48867	0.488778	0.777682	0.82664	0.409884	1.11142	0.6592
0.5	0.972539	0.646156	1.01175	0.453035	1.76429	0.775927	1.45876	0.658409	1.41051	0.304161
0.6	0.492826	0.692448	0.59595	0.55683	0.970373	0.660768	1.424	1.0292	1.11701	1.49548
0.7	0.713656	1.09715	0.54967	0.55814	1.9028	1.91539	1.82231	0.435818	1.39446	0.576588
0.8	0.772517	0.666612	0.720153	1.71004	1.14163	0.697635	1.14189	1.41322	1.73418	0.894949
0.9	0.993887	0.539212	1.23366	1.70034	0.516572	0.74322	0.488951	0.846514	1.46258	0.576887
1.	2.20376	0.875916	0.522533	2.10097	1.50679	0.765368	0.818627	1.60007	1.53174	1.39404

Table 5
TELL: *p*-values of the *T* statistics for $k=2$ and $q=10$

ρ_1 / ρ_2	0.1	0.2	0.3	0.4	0.5	0.6	0.7	0.8	0.9	1.
0.1	0.47428	0.55162	0.31822	0.72244	0.54935	0.81984	0.73891	0.52179	0.6813	0.85669
0.2	0.97159	0.05434	0.64401	0.36426	0.06822	0.59059	0.34787	0.33832	0.23839	0.66318
0.3	0.4831	0.57844	0.23259	0.11305	0.4386	0.72928	0.96781	0.06237	0.08854	0.40753
0.4	0.79968	0.47713	0.89013	0.14285	0.90616	0.59544	0.54632	0.9644	0.31675	0.72462
0.5	0.41632	0.7389	0.3858	0.93572	0.07868	0.597	0.15214	0.72543	0.16887	0.99728
0.6	0.9025	0.6872	0.79596	0.83855	0.41807	0.72291	0.16416	0.37283	0.31311	0.14088
0.7	0.66383	0.32648	0.8458	0.83709	0.05797	0.05645	0.06897	0.9484	0.17472	0.81785
0.8	0.60045	0.71635	0.65679	0.0882	0.2981	0.6816	0.29795	0.16789	0.08389	0.48252
0.9	0.39946	0.85665	0.24631	0.08996	0.87938	0.63191	0.90601	0.52731	0.15099	0.81753
1.	0.02912	0.49993	0.87357	0.03713	0.13772	0.60823	0.55468	0.11164	0.12982	0.17489

Across ρ_1 the mean values of *p*-values are: 0.574033, 0.548704, 0.588918, 0.416925, 0.383225, 0.603325, 0.474482, 0.474038, 0.234638, 0.608297. As before still we cannot reject normality.

References

1. Zinger AA (1956) On a problem of A. N. Kolmogorov. Vestnik Leningradskogo Universiteta 1:53–56
2. Sakata T (1977) A test of normality based on some characterization theorem. Mem Fac Sci Kyushu Univ Ser A 31:221–225
3. Sakata T (1977) Two characterization theorems of normal density function. Mem Fac Sci Kyushu Univ Ser A 31:215–219
4. Bakshaev A (2010) N-distance test of uniformity on the hypersphere. Nonlinear Analysis: Modelling and Control 15:15–28
5. Klebanov LB (2005) N-distances and their applications. Karolinum Press, Prague

Index

Andrei Y. Yakovlev et al. (eds.), *Statistical Methods for Microarray Data Analysis: Methods and Protocols*,
Methods in Molecular Biology, vol. 972, DOI 10.1007/978-1-60327-337-4, © Springer Science+Business Media New York 2013

M

N

O

P

Q

R

S

Lightning Source UK Ltd.
Milton Keynes UK
UKOW07n0602260516

274973UK00005B/83/P